SHARING SPACES

INTERSECTIONS: HISTORIES OF ENVIRONMENT, SCIENCE, AND TECHNOLOGY IN THE ANTHROPOCENE

SARAH ELKIND AND FINN ARNE JØRGENSEN, EDITORS

SHARING SPACES

TECHNOLOGY, MEDIATION, AND HUMAN-ANIMAL RELATIONSHIPS

EDITED BY   FINN ARNE JØRGENSEN AND DOLLY JØRGENSEN

UNIVERSITY OF PITTSBURGH PRESS

Published by the University of Pittsburgh Press, Pittsburgh, Pa., 15260
Copyright © 2024, University of Pittsburgh Press
All rights reserved
Manufactured in the United States of America
Printed on acid-free paper
10 9 8 7 6 5 4 3 2 1

Cataloging-in-Publication data is available from the Library of Congress

ISBN 13: 978-0-8229-4830-8
ISBN 10: 0-8229-4830-3

Cover photo: Courtesy of the US National Park Service trail cameras.
Cover design: Alex Wolfe

CONTENTS

SHARING SPACES

BEARING WITNESS TO TECHNOLOGICAL MULTISPECIES ENTANGLEMENTS

FINN ARNE JØRGENSEN AND DOLLY JØRGENSEN

A grizzly bear, brown and powerful, is frantically trying to escape a metal wire snare among some timber logs in the woods. We watch as rangers arrive, load a gun with a dart, and sedate the bear. They examine her (she is three years old, in good shape, big teeth), equip her with a VHF collar, and tag her ear with a yellow plastic label numbered "71." After the bear wakes up, they release her, using fireworks to scare her away and condition her to avoid humans. From this point on, she is no longer a fully wild bear, doing whatever bears do in the woods. She is monitored and managed for the rest of her life.

Bear 71 was the protagonist of an award-winning interactive documentary released by the National Film Board of Canada (NFB) in 2012. The twenty-minute installation, viewable as a Flash-based website or as a virtual reality installation, follows the life of this one particular bear in Banff National Park from when she was captured, collared, and tagged at three years old until her death eleven years later.[1] Using a wide range of data collected in the Banff National Park, the team behind the documentary, headed by Leanne Allison and Jeremy Mende, could follow this bear as an individual in order to tell us not only something about nature, but also something about technology as a mediator and connector of lives across time and space.

On the website, after the short video introduction where we watch Bear 71's capture, we are presented with a somewhat abstract and stylized 3-D map of a natural landscape, with dots and symbols moving around. The viewer is represented by an icon on the map, and we have the freedom to move around the map to explore. At times a scripted presentation takes over, with narration, text, graphics, and video. In wandering around the map, we encounter a wide range of animals, each of them tagged with a number: not only Bear 71, but also other grizzly bears, black bears, deer, wolves, cougars, ravens, lynxes, crows, hares, big horn sheep, foxes, elks, mergansers, eagles, coyotes, dogs, humans, and horses. Some will wander in from beyond the edge of the map, some will leave the area, while others remain. Each has been observed at some point by a camera trap or other sensor, and clicking the icon of each creature brings up a video recorded by the camera trap. We watch them move around in the landscape; observe how the paths of people, animals, and things like cars and trains cross; and notice how alive this place is with living beings that move around and share the same place.

People, animals, infrastructure, trains, and cars inhabit, comprise, and move around this landscape. On the map interface through which we access the landscape, we see how multiple species share a space and inhabit a joint world, but are often unaware of each other. The video footage from the trail cams illustrates this well. For instance, on one hiking trail obviously made by people for human purposes, a bear comes up and smells around the area. Later, we see a moose walk across the same trail. Another trail cam is located at the underpass of a freeway where we can see a bear, a moose, and a group of people with inflatable bathing toys cross at different times. Yet another trail cam is located by the train tracks, where we can see a grain spill from a train. After a few minutes, a bear comes out of forest to eat grain. This is not an uncommon scene. A pop-up text mentions how the railroad has spent tens of millions of dollars trying to reduce animal deaths on tracks. Later, during one of the scripted segments of the documentary, we learn that this is how Bear 71 died—run over by a train while eating grain off the tracks. In bringing together all available data, Bear 71 comes to life—and death—for the viewers. While one can discuss the appropriateness of the anthropomorphic narration that the documentary attributes to Bear 71, the bear does become an individual through the myriad traces left in technological media.

Environmental historian Tina Loo has likewise tracked the life of an individual bear in Banff National Park.[2] Bear 148 was a well-known figure in the Bow Valley, having many documented encounters with people. However, these encounters eventually became too many, too close, and too threatening, so she was relocated to a more remote area. As Loo states, "We don't think of wild animals like grizzly bears as having a history, but they do, and not just collectively and evolutionarily, but as individuals." Technologies of monitoring allow people to get to know these animals as individuals. Together with artist and grizzly bear biologist Colleen Campbell, Loo explored location data as a way of writing animal history.[3] Building on artist Aaron Koblin's observation that movement makes space, Campbell and Loo show how bear location data gathered by researchers give them some insight into bear country—landscapes as used and experienced by bears.[4] Yet writing animal history based on such data requires a practice of critical empathy, informed by a deep understanding of bear behavior and environments.

A similar phenomenon takes place every year when the live streaming Bearcam from Brooks Falls, Alaska, starts up, and hundreds of thousands of viewers watch bears fatten up on salmon in the river. The web page hosting the live stream coming in from Katmai National Park and Preserve in Alaska has hundreds of thousands of comments where people discuss the bears, which have names like Chunk, Grazer, Otis, Holly, and Riffles in addition to their numeric identifiers. The webcam creates a real connection between watchers and bears: "When I sit in front of the computer, along with thousands of other viewers, mesmerized by the bears of Brooks Falls catching salmon, the bears and I are close together, despite being physically far apart."[5]

Yet another bear, Bear JJ1, more commonly known as Bruno in the German media, serves as a counterpoint to this story.[6] Bear JJ1 was born in northern Italy, but made history when he wandered across the border to Austria and eventually to Germany in 2006.[7] Unlike Bears 71 and 148, JJ1 did not have a radio collar and could not be monitored in the same way. He was first observed in early May close to the Austrian village Tösens and would in the weeks to come raid farms and kill some farm animals, typically without eating them. Bear JJ1 evaded capture and control, however. Most observations were after the fact, consisting of dead animals, droppings, and the occasional sighting and subsequent reporting by locals.

The Austrian Bear Emergency Team, which had been founded to intervene with and manage wild bears in Austria, initially could not even reliably identify the bear and where it came from—was it one of Austria's small native population or had it wandered in from elsewhere? JJ1's behavior raised serious concern, as he seemed to kill animals for sport and showed little fear of humans. Austrian officials launched a plan to capture and radio-collar JJ1 so that they could anticipate his movements and intervene more easily should there be more problems. However, before they got that far, JJ1 left Austria and wandered into Germany, the first bear there in 150 years. JJ1 continued his killing spree in Germany, eventually triggering massive public debate about his fate. JJ1 became both a media phenomenon, getting the anthropomorphic moniker "Bruno," and a political problem, being called a "problem bear." Humans pass judgment on animal behavior. In the end Bruno was shot on June 26. Many argued that JJ1 should instead have been anesthetized and equipped with a GPS collar, in the belief that this would have brought him under control. Without such tracking, he died as he lived: unruly and unmanaged, challenging human valuations.

Bear 71, Bear 148, Bear JJ1, and the bears of Brooks Falls come across as individuals with individual stories. Their lives generate interest and their deaths become tragedies—they feel personal. However, it is only through their embedding in technological management systems, through monitoring and tagging, that we get to know them.[8] The technology of tracking and visualizing these data allows humans to see this shared space and recognize that it exists. Technology has become a key part of how we know animals. Technology is mobilized to give us knowledge about a particular animal, but it also structures the relationship we humans have with that animal.

This book, which originated as a series of papers at a workshop at the Nordic Centre at Fudan University in Shanghai, China, explores precisely this tension between enabling and structuring human-animal relationships across space.[9] We seek to understand how technologies are part of the fabric that connect humans and animals to each other and the spaces we share.

SHARING

Humans do not live alone in the world; they are always in relation with other species, from bears to bacteria. Humanities and social science scholars have begun to investigate these relationships, their reciprocity and their tensions, under the loose name "multispecies studies," a formation that builds on decades of work in animal history, feminist studies, and Indigenous epistemologies. Multispecies studies focuses on "the multitudes of lively agents that bring one another into being through entangled relations."[10] This is the kind of entanglement we see in the bear stories: animals' lives intersect with the lives of others, whether through direct physical contact or even inhabiting the same space at a different time.

Bruno Latour has argued that the world is and always has been a place of socio-natural hybrids—"naturecultures," which are a blending of environment and technology.[11] Natureculture as an epistemological, ethical, and aesthetic category has been widely accepted in the humanities and social sciences as a condition of the Anthropocene. With a multispecies approach, naturecultures extend beyond the technological to encompass how the other species around us are bound up with human cultures. The entanglement of species demonstrates that the multispecies world that we inhabit is by necessity a hybrid natureculture.[12] While this kind of entanglement is easy to see (though not actually easy to unravel analytically) with domesticated species,[13] it exists just as strongly with wild animals. Bears like 71 and JJ1, the people who collared and tracked them, and the geotracking technologies deployed together create a hybrid natureculture world.

Scholars in the multispecies studies field have pushed for an "attentiveness" to relationality and the ethics of entanglement.[14] Because there has never been a moment in human history in which animals did not play a part, writing animals out of the narrative "is a methodological, and even political choice" that needs to be countered.[15] Multispecies scholars advocate "staying with the trouble" to focus on the complexity of multispecies relations.[16] There are multiple ways of knowing and defining an animal, which lead to human-animal interaction as an "active site of engagement," as Zoe Todd has theorized with the case of human-fish relations in arctic Canada.[17] Different actor groups can enact what an animal is and how it should/could be related to in diametrically opposed ways.[18] The reciprocal relations between human and animal underpin complex political, cultural, and social realities.

SPACE

Throughout this volume, we encounter humans and animals sharing a wide range of different spaces: in forests, in cities, underwater, and in structures like laboratories, farms, and abattoirs. These spaces emerge from interactions between humans and animals. We build on Doreen Massey's understanding of space as "the product of interrelations, as constituted through interactions," and thus "always under construction."[19] The spaces we share are heterotopic, overlapping, and fluctuating, ever dynamic and flowing. They are subjectively experienced by people and animals alike, simultaneously existing "out there" in the world and inside the minds of people and animals, shared and yet not shared, or what Edward Soja calls "real-and-imagined" spaces.[20] They are lines on maps that can be mobilized for a variety of purposes. They are filled with natural, cultural, and infrastructural features.

Following the so-called spatial turn, scholars in a wide range of fields integrated space into their studies, exploring the many ways in which spaces comprise the world.[21] Geographers in particular have been concerned with the where-ness of the world,[22] but fields like environmental history have also been intensely interested in the spatial character of historical events.[23] For instance, Bill Cronon's hugely influential *Nature's Metropolis* demonstrates how a metropolis like Chicago creates an extensive set of spatial relations around itself.[24] Cultural ideas are attached to space and place as well.

The bears discussed in this chapter, as well all the other animals in this book, direct our attention to the spatial relationships of humans and animals. Space is the fabric of all narrative and all interaction. History takes place here. As Øyvind Eide has argued, "People move in space, and we live our lives in time."[25] Both humans and animals live their lives in time and space, always coexisting in either one or both. When they coexist in both, they physically encounter each other; when they coexist in only one, they may or may not be aware of each other.

Animals have made themselves at home in "our" space, because in fact it is not ours.[26] Chris Philo and Chris Wilbert argued in their volume *Animal Spaces, Beastly Places* that human ideas of animal spaces, where animals belong and what they should be doing there, are different from beastly places inhabited and used by animals for their own purposes.[27] Their book was a significant contribution to the field of animal geography, making a strong claim for why social science needs to consider animals. Philo and Wilbert are central in the third wave of animal geography, defined by Julie Urbanik, as an interdisciplinary and multidisciplinary effort "to see other-than-human beings as actors in the world."[28] There is now a recognition that human-animal interspecies relations are located in space and geography.[29]

How animals are monitored, interpreted, and categorized has spatial implications for them. Like we do in this chapter, Hodgetts and Lorimer open their exploration of animal spaces with bears, comparing wild-roaming and sometimes transgressing polar bears with pacing and depressed polar bears in captivity.[30] Space matters to these bears. As the Bear 71 installation demonstrates, the space portrayed is defined by the bear, rather than being a human definition of space. We share space, but that space may not be defined or constructed in the same way by all the parties involved. Power relations are at work in the making of space. Following Lefevbre, we realize that space is not an empty container or simply a place where people and animals meet. Rather, space emerges from the interactions we have with animals. Space is "a medium through which social relations are produced and reproduced."[31]

TECHNOLOGY, MEDIATION, AND HUMAN-ANIMAL RELATIONSHIPS

In this volume, we argue that sharing space between animals and humans is structured by technology. Technology, as we define it, is not restricted to modern digital technology—everything humans create as tools, from clothing to cars to computers, is technology. In the last decade it has become commonplace to attribute explanatory power to digital technology as a negative force in the mediated human relationship to the world;[32] however, this approach lacks historical insight—analog technologies from aquarium glass to paper maps also mediate the human-nature bond. Technologies, whether digital or analog, are neither good nor bad, nor are they neutral.[33]

Scholars in the field of science and technology studies have examined technology as a mediator. As Latour has defined them, mediators "transform, translate, distort, and modify the meaning or the elements they are supposed to carry."[34] Mediators are different from intermediaries, which do not transform the things they carry but simply transmit them as whole packages. Mediation can affect both the physical thing (whether animal, person, building, or object) and the perception of that thing. In this volume we show that technologies mediate human-animal relations.

The relationship between humans and animals takes place in particular spaces, so the technologies that create those shared spaces mediate the relations. The roads, railroads, and hiking trails in the bear stories all structure the places where meetings can happen. Some technologies, like the highway underpasses, disconnect animal mobilities from the human automobile traffic, whereas grain spilled along the railroad tracks attracts bears to a dangerous meeting point. Camera traps capture images of humans and animals alike traversing the landscape, and digital technologies make those images available to others at a distance. Technological mediation can decouple spatial closeness from physical closeness.

Seeing images can give a sense of proximity, but what kind of encounters are possible when the historically and spatially situated context of the animal (and human) moves out of view? As Zeb Tortorici argues, based on a stereoscope card of Kodiak bears in the Brookfield Zoo, "exposure and observation of animals in captivity and death paradoxically distances the human viewer, both visually and ontologically, from the animal that is being viewed."[35] On the stereograph these bears have a three-dimensional lively quality, yet what is shown is a scene of captivity and suffering. Even in the case of Bear 71, there is a question of whether or not capturing and displaying the bear's death create a similar feeling of distance. The joggers who run in the same area are far less likely to be killed by a train, which demonstrates the fundamental power imbalance between these species that share space.

Humans have deliberately modified animals through domestication processes in order to fit them into our technological systems, while those systems are also modified around those animals in order to function. Animals from hunting dogs to work horses to dairy cows are technologically created beings modified by both breeding practices and the technological artifacts hooked up to them to convert their energy into work and products for humans.[36] It might be easy to see how technology is used in the automated milking shed with domesticated cows, but even wild animals are "domesticated" in a sense under technological regimes: they are collared, tracked, sedated, moved, sampled, and more.[37]

Technology is used for control, but also for care and conservation. Genetic technology, for example, is being used to re-create extinct species, a process labeled de-extinction. Projects to revive the passenger pigeon, mammoth, and thylacine are all ongoing. Such a time- and money-intensive technological process has been motivated by a desire to restore species and ecosystems, as well as a sense of guilt and grief for humans having destroyed them in the first place.[38] While advocates argue that the end result of this technological intervention is animal conservation, others have pointed to ethical problems, including the high number of individual animals who die during the technological trials, the uncertain availability of appropriate habitat for re-created species, and the funneling of funding to what might be seen as vanity projects.[39] As we wrote earlier, referencing Kranzberg's laws, technology is neither good nor bad, nor is it neutral.

OUR SHARED SPACES

This book explores nonhuman animals as kin, as companions, as food, as transgressor, as entertainment, and as tool. The relationships between those animals and humans happen in space through the mediation of technology. Technologies create the spaces of production, consumption, labor, and contact that bring humans and nonhumans into relationship.

We start with three chapters about domesticated animals. Nicole Welk-Joerger explores the calorimeter as a technological site at which human-animal relationships changed. The calorimeter was designed to scientifically measure the conversion of cattle feed into human food through the body of the animal. But the technological design required humans and cattle to work cooperatively, prompting the scientists to select particularly docile animals and certain feed mixes. The technological choice of the calorimeter as the space of interaction reinforced industrial thinking about human-animal relations on the farm.

Tatsuya Mitsuda takes us into the slaughterhouse in the second chapter. Using the case of occupied Qingdao at the beginning of the twentieth century, he investigates how Chinese cattle bodies were envisioned and processed in order to meet the transnational demand for milk and beef. He discusses how these bodies were "made fit for" particular human populations through hygienic and scientific techniques.

Based on ethnographic fieldwork, Aurore Dumont explores how technology shapes the relationship between pastoralists and their herds in contemporary Inner Mongolia, People's Republic of China. She argues that many anthropologists have discounted technology among these Tungus and Mongol pastoralists, writing about it as material culture or technique, even though the use of modern technologies such as motorized vehicles, mobile phones, and GPS abounds in these societies. The introduction of mechanized mobility technologies because of structural and policy changes in the latter half of the twentieth century has created new patterns: previously nomadic herders have moved from traditional yurts into mobile homes and travel by truck, which allows settlement farther from herds and reduced physical interaction with the animals. Technological change has modified the spatial relationship of pastoralist and herd.

Human-animal entanglements through hunting and scientific conservation are also technologically mediated. Karin Dirke offers a close reading of hunting stories in the Swedish hunting magazine *Jägaren* at the turn of the twentieth century to illustrate how the killing of animals was managed through the entanglement of ethics and technology. She shows that the inherent violence of the hunt was ameliorated in the minds of Swedish hunters through the deployment of technology that would kill the animals as quickly and "cleanly" as possible. The magazine itself also serves as a space of encounter for these hunting stories, with language evoking the soundscape and chronography of the scene in order to allow hunters to experience another hunter's remembrance of space and animal encounters.

The next chapter continues the theme of hunting technology and animals, but brings it to contemporary times. Finn Arne Jørgensen examines the use of GPS collars on dogs used during moose hunting in Sweden. Although hunting has always relied on tools, the integration of GPS tracking collars in hunting practices augments the hunter's ability to remotely sense the environment. Jørgensen finds that the triad relationship of hunter-dog-moose is mediated through the GPS device: the hunter knows where his dog is through the screen and in turn can "see" the moose on the screen by interpreting the dog's movements.

Heta Lähdesmäki's chapter turns to a different but related use of tracking collars: wolf conservation in Finland. She shows how the human-wolf interaction is co-constructed through the collar data. Radio-tracking practices require humans and wolves to have physical contact when the animal is caught for collaring, but then permit researchers to be close at a distance through the data signal. Through GPS signals, wolves became individuals with spatial and temporal histories, distinct from the wolves of myth and folklore.

We stay in the fields of Finland with an investigation by Tuomas Räsänen into human-eagle relations. Hated, feared, and often killed by legal and illegal hunting in the early 1900s, and threatened by environmental toxins in the postwar years, white-tailed eagles were almost extinct around the Baltic Sea by the 1970s. He argues that Finnish and Swedish conservationists rescued the eagles through a new type of "care protection" for birds, where the nurturing of individual animals was central to the success of the program. Technology mediated relationships, both between conservationists and between people and eagles, in ways that both hindered and promoted conservation.

In Ellen Arnold's chapter, we face an unexpected question: How do you send penguins by mail? This question arose when, in the 1930s, penguins arrived in American zoos for the first time in the wake of Richard E. Byrd's Antarctic expeditions. Trying to raise funds for future expeditions, Byrd offered twenty penguins for sale. In the deluge of mail from individuals who wanted to purchase penguins, the issue of how they would be transported was a central question. Arnold argues that transportation and communications technologies served to put penguins in the American environmental imagination.

Technologically mediated human-animal encounters also affect how and when we "see" animals. Dolly Jørgensen's chapter demonstrates how aquariums are planned and constructed spaces for visual encounters with aquatic species. Humans have looked at fish for thousands of years, but the aquarium made possible new ways of looking at aquatic life. Developments in material technology fundamentally changed the character of spectatorship interfaces. She shows how such mediating technologies connect, but also separate and distort, human-animal relationships.

Charity Edwards and Amelia Hine move us deeper underwater. Through their academic design backgrounds, they explore a speculative space in which southern elephant seals are used to remotely conduct oceanographic mapping and experimentation. The elephant seals are outfitted with technological devices that allow the collection of deep-sea data on their regular feeding dives. These data are then put into the service of the extraction industry. The animals and their counterpart, autonomous underwater vehicles, converge as fellow workers in seabed mining. In Edwards and Hine's reading of the work of these elephant seals, the seabed takes on urban logics as the wild is domesticated.

Riin Magnus explores the use of electronic mobility aids in the communication between blind people and their guide dogs. As she presents it, every assistive device allows for different *Umwelten* for its users, building on Jacob von Uexküll's terminology for understanding the relationship between the self and the world. Tracing the development of such mobility aids back to World War I, Magnus demonstrates how the training of guide dogs combined with new technologies to enable communication between humans and dogs.

In the final chapter Concepción Cortés Zulueta explores what she calls a "technological love triangle" between humans, cameras, and other animals. She argues that cameras have changed the emotional framing of human encounters with other animals, reducing the perceived distance between them in significant ways. While pre-camera encounters typically were framed by a "Freeze, Flight, or Fight" response (in other words, considering animals as a perceived threat), Cortés Zulueta argues that a "Picture, Pet, and Play" response has become much more common.

To round off the volume Jenny L. Smith offers her reflections on the potential of scholarly inquiry into the technological mediation of human-animal relations. Noting the asymmetry of the sharing of space and the exploitation of animals through technology, she cautions that in the stories in this volume humans have often failed to consider their responsibilities to the animals with which they are entangled. She also finds visibility as a strand running through the papers in the forms of tracking, accounting, and display. While technology can make animals visible, scholars have a responsibility to make visible the entanglements created by the process.

The essays in this volume demonstrate how in modern society we have witnessed the rise of many types of technological ways of *being with animals*. Technologies at all scales, from the personal to the infrastructural, have become increasingly embedded in both natural and human-made environments since the mid-nineteenth century. The nonhuman natures that humans encounter have become thoroughly entangled with human technologies and societies. Transmission collars on bears, dogs, and wolves, as well as cameras, computers, and trains, all provide new ways of interacting with nonhuman nature. Through standardization, mechanization, and digitization, animals have become part of human machines. We encounter the bears of Brooks Falls and Bear 71 through our home computers and mobile phones, themselves just minor components of our networked digital society and its system of machines. We are *with animals* through technological means.

Now we as scholars in the environmental humanities and affiliated fields are called to witness how ways of looking at, measuring, moving, and killing, as well as controlling, containing, conserving, and cooperating with, animals have shaped human relationships with the nonhuman world. Multispecies scholars have urged us to give attention to human-animal entanglements; as scholars we have a responsibility to bear witness to the place of animals in our lives—and our place in animals' lives. We all share space and stories. Technology has to be an integrated part of those stories.

CHAPTER 1

COOPERATIVE CALORIMETRY AND THE INDUSTRIALIZATION OF CATTLE FEEDING

NICOLE WELK-JOERGER

In 1901 scientists at Pennsylvania State College placed a steer on a strict diet. For fourteen days, he gobbled up prepared rations of timothy hay with some added linseed meal. In the first twelve days of this diet, he spent his time in a small stall, to which he acclimated as his primary living space. The scientists took great care in weighing his food before placing it into his feed box. As he ate, the steer wore a special harness around his belly so that his urine could be meticulously captured, measured, and analyzed for its various chemical properties.[1] For the last two days of his carefully monitored feeding schedule, the steer was moved from his small stall to a large box-shaped device located across the building: the Armsby respiration calorimeter.

Scientists walked the steer up a narrow ramp leading into the device. They secured the heavy oak door—24 cm thick with wood, metal wires, and an airlock seal—behind him. The machine hummed as air moved in and out of the chamber.[2] Mirroring his other stall, the steer lived in this artificial environment for forty-eight hours. It was rumored he wouldn't be able to hear or see much from inside this box, but gushing water, clanks, and clicks could be distinguished from the murmurs that radiated softly through a small window at the left side of his head.[3] The scientists watched the steer through this window and logged his activity. He stood, and he lay down. He ate and drank from a specially made feed box that allowed the scientists to give him his ration without entering the device. The steer's mood even changed from "calm" to "anxious," from chewing his cud to responding to a foreign noise.[4] All the while, the steer's intake, excrement, air, and heat changed during these periods and were measured by the scientists.

FIGURE 1.1. A steer wearing a harness that enabled scientists to collect and measure urine at Pennsylvania State College. Undated [likely 1915]. *Source*: Eberly Family Special Collections Library, Pennsylvania State University.

Whether the steer realized it or not, his body provided important clues for understanding how feed was converted into meat and milk by food-producing bovines like himself. At that time, although farmers knew *what* to feed cattle, it was unclear *how* this material was used in the body. Some feed was converted into heat that was expelled from the steer. Other parts contributed directly to meat production, with unprocessed material turned into waste. The data obtained from the steer's stay in the calorimeter were published by the US Department of Agriculture's (USDA) Bureau of Animal Industry and circulated widely for scientists, regulators, and farmers to read. The studies from this steer, among other bovines who entered the Armsby calorimeter, suggested that increased meat production relied on the addition and careful curation of supplemental feed in cattle diets.

These conclusions were formed at the height of the United States' Progressive Era, a period marked by public activism and political reform in many arenas, including food production and consumption. During this time in the late nineteenth and early twentieth centuries, Americans worried about their food, especially about the integrity and purity of their food products in the wake of increasingly industrialized food companies. The moral prerogative of food safety and food security dominated the psyche of the scientists, regulators, and temperance leaders of this time, and would go on to inform the development of new ways to define and measure food.[5] Measurement took special precedence in conversations about ever-looming scarcity and consistently pursued growth, with the efficacy of nutrition, in its capacity to sustain health and labor in a capitalist system, often reduced to production gains.

Although the story of Progressive Era food reform has primarily focused on human food products, sources make clear that the food fed to animals was of equal concern in the eyes of regulatory and scientific leaders. Toxic and poisoned animal feed often made headlines in newspapers,[6] feed was a point of discussion during Pure Food Meetings and on federal trade boards,[7] and the 1906 Pure Food and Drug Act was drafted by leading agricultural chemists to simultaneously protect human food and animal feed.[8] To safeguard the integrity of America's large food system, these leaders could not afford to ignore what and how their domesticated animals were fed. This included attending to both safety and efficacy in livestock feeding, and scientists, with funding from policymakers, dedicated their time and resources to these efforts.

Attending to questions of nutrients, efficacy, and gains, in 1898 Pennsylvania State College scientist Henry Prentiss Armsby proposed building a livestock-size calorimeter to measure the energy values of animal feed. At the time nutritive values were being calculated for federal guidelines in human food using a respiration calorimeter designed by Wilbur Atwater, who conducted his experiments at Wesleyan University. The purpose of Armsby's device was to understand how feed was processed in animal bodies in order to better calculate what would later be called a "balanced ration": the most efficient balance of different feeds that encouraged high meat and milk production. Pennsylvania State College's calorimeter was eventually approved and funded by the Bureau of Animal Industry, which solidified its status as a national tool meant to inform federal studies and regulations.[9]

The first animals tested inside Armsby's calorimeter were cattle—an unsurprising decision (with an accompanying technological design) given the significance of bovines as sources of both food and political power in the United States. Cattle had aided in early European colonization efforts, and as moving property, they continued to help white settlers dispossess land and resources from Indigenous groups in nineteenth-century westward movement.[10] During this same period social Darwinist frames of thinking influenced attitudes about the inclusion of beef and milk in American diets. Beef signaled both racial and social elitism in Progressive Era circles, privileging both whiteness and wealth. Access to meat and the consumption of beef anchored conversations about social uplift, with beef cited as a source of intellectual and physical prowess.[11] Cow's milk, relatedly, served as an important basis in conversations about food purity and vitality.[12] Thus both beef- and milk-producing bovines occupied important places in the American imagination and stomach, concretizing the significance of cattle as major players in scientific attempts to standardize (and, in turn, increase) national meat production.

Before the calorimeter, farmers and scientists used carefully formed menus and isolated feeding stalls to test how feed impacted animal production. They obtained estimated results through incremental ration building and elimination diets. These interactions between the researchers and their animals echoed those produced on the farm. In fact, many earlier agricultural improvers conducted experiments on their own herds using their own resources and larger farming operations.[13] With government investment in agricultural research, beginning with the Morrill Act of 1862, formally trained scientists replicated similar experiments on state-owned farms. However, the reputation of the laboratory as an objective space required the adoption and use of instruments of study otherwise inaccessible to the average farmer. In the Progressive Era the use of cutting-edge tools in the reform and improvement of sectors like health, food, and agriculture became important markers of expertise. In animal feeding the calorimeter partially served the purpose of representing this supposed intellectual hierarchy.

The use of this enclosed, federally funded technological instrument transformed how humans and cattle worked with one another in the lab and, eventually, on the farm. These changes helped form the foundation of twentieth-century industrialized food animal production, in part because calorimeter results rendered the inner workings of livestock as measurable factories, able to be modified by scientists and, by extension, farmers. To achieve the standardized conclusions needed to formulate their agricultural recommendations, scientists applied an industrial logic for agriculture to animal bodies.

Applying this logic and using a special technological instrument to reach these conclusions certainly did not come naturally. As scholars, including Daniel Todes, Karen Rader, and Tiago Saraiva, have shown, standardized animals are *made* through various political and professional processes.[14] The bovine participants of the calorimeter experiments are no exception. To form their universal feeding recommendations, scientists relied specifically on cooperative cattle: laboratory animals who worked in hyper-controlled, confined spaces. The novelty of the device forced researchers to tinker and troubleshoot solutions as they ran into problems that had been largely nonexistent in earlier feeding experiments. Animals were expected to be trainable and amicable, replicating seemingly everyday behaviors in a fully enclosed foreign box that made strange noises. Further, these cattle needed to adjust to the various inventions, some even strapped to their bodies, that aided scientists in the collection of their bodily inputs and outputs. The calorimeter was thus a technological site where researchers and their animals came together in close physical and intellectual proximity to make sense of a theoretically complex aspect of food production and to better standardize—and industrialize—feed and feeding.

The recommendations the lab made ultimately privileged a particular kind of relationship humans could have with domesticated food animals, suggesting what researchers needed from their experimental animals was similar to the needs of farmers from their cattle. The industrial logic scientists applied to cattle in the calorimeter lab diffused across agricultural communities through extension services and agribusinesses. Overall, the human-animal relationships fostered at the lab formed the foundation for later twentieth-century developments in industrialized food animal production, reliant on a network of experts and outsourced products that enabled the later expansion of farm operations.

This story of the calorimeter complicates two features that emerge in histories of nonhuman animals and the technologies at times built on or around them. First, it demonstrates how humans designed technologies to measure animals with specific assumptions about their behavior in mind. Scholars in the history of science and technology have demonstrated how societal assumptions (patriarchy and masculinity, white supremacy, definitions of democracy) have long informed human observations of nature.[15] In the more specific case of the calorimeter, assumptions about the role of cattle in food production—and the related pursuit to control and manage their bodies—informed the technological design, laboratory interactions, and later the farm interactions humans had with their animals. Second, even as scientists observed cattle inside technologies, which arguably further rendered their bodies as mechanized, "industrialized" organisms,[16] cattle continued to demonstrate their individuality in laboratories and on farms. These behaviors, expressed as *both* resistance *and* compliance to confinement, contributed to attentions and attitudes about food animal industrialization. These attitudes continue to be grappled with by farmers today who work with standardizable but also very individualistic bovines on industrial farming operations. The tension between the two did not disappear with the use of an "objective" technological apparatus, even as its creation and use were meant to meticulously measure a universal figure of feeding in the pursuit of more beef and milk in American diets.

A CALORIMETER FOR CATTLE

Throughout 1898 Henry Armsby corresponded with Wilbur Atwater about the materials required to enlarge Atwater's calorimeter model. Armsby was specifically concerned about available refrigeration technologies and the electrical power needed to operate these parts of a larger calorimeter. In budgeting the expenses for such a device, he was left estimating its cost using rough figures from Atwater's device and comparing these with the operating expenses of the new refrigeration technologies in Penn State's updated creamery.[17] In the end it cost approximately $20,000 to construct the device and house it in its own building.[18]

FIGURE 1.2. A later image of the Armsby calorimeter, pictured in 1955 with researcher Raymond Swift. *Source*: Pennsylvania State University Eberly Family Special Collections Library.

Before its official completion in 1902, the livestock-size calorimeter was subject to a national media hype, especially as it related to animal welfare. In addition to its status as a new and "peculiar" instrument,[19] its timing was concurrent with the formation of groups advocating for the welfare of experimental animals. In 1883 the American Anti-Vivisection Society was established in Philadelphia, with similar groups emerging in Boston in 1895, Baltimore in 1898, and Chicago in 1903.[20] Much like the American Society for the Prevention of Cruelty to Animals founded before them, these organizations advocated for the welfare of small animals and livestock, particularly for improved conditions for cattle transported across the country via railcar. Newspapers played to the ethical concerns of these groups, and with the emergence of the calorimeter the question of livestock experimentation was certainly on the table.

Compared to other forms of experimentation, the calorimeter was a different kind of experimental instrument. The device was able to measure the inner workings of the animals without cutting into their bodies. One *Chicago Tribune* article showcased this novelty with emphasis on the possible (or hopeful) response from antivivisectionists: "Anti-vivisectionists will have no cause to throw up their hands in horror. The beasts are not to be smothered or otherwise slain, tortured, dissected, or made uncomfortable in any way. Fed upon the choicest foods and housed in ideally comfortable quarters, always at an agreeable and uniform temperature, they will be the happiest beasts alive." Calorimeter experimental subjects were to be considered some of the "happiest beasts," even if they were "imprisoned" in an "airtight cell," as the article's title suggested to catch their readers' attention.[21] Media coverage reflected a unique tension that was building in contemporaneous animal welfare debates three years before Britain's torrid "Brown Dog Affair,"[22] with the *Tribune* article weighing the respected avoidance of vivisection with the potential discomfort for any animal to be sealed inside a layered metal box.

Technically speaking, and further emphasized by these newspapers, Armsby's calorimeter was the first of its kind in the world. It was large enough to hold a steer, or a man milking a cow inside its chamber.[23] Unlike its large-scale predecessor, which Armsby visited at the Möckern Experiment Station in Leipzig, Germany,[24] his device was both a direct and indirect calorimeter. It was able to measure body heat as well as the oxygen inhaled and carbon dioxide exhaled from the animal inside it. This was based on Atwater's novel construction.[25] The many-layered box, made with both wood and metal, accounted for these gases and energy expelled from the animal's body. The scientists meticulously measured the copper tubes of cool water running along all sides of its walls. Temperature fluctuations at a fraction of a degree signaled burned calories.[26] These calories helped calculate how much energy from a given feed was used by an animal's body to produce meat and milk for humans.

The calorimeter required a dedicated team of researchers to operate. At least one person was required to monitor all forty-eight to seventy-two hours of a given experimental period and log six readings of the device every four minutes. A master clock was used to signal the timing of these measurements as well as shift rotations. Ideally, two researchers oversaw the calorimeter at the same time to cover alternating thirty-minute shifts.[27] The calorimeter attracted new students and high-caliber researchers interested in working with Armsby and the device that constituted the US Institute of Animal Nutrition. These students were largely men, but the team also included some women, who worked as librarians, editors, and stenographers. These women were instrumental members of the research team, and it is likely they aided in taking down the four-minute measurements in addition to taking shorthand notes, collecting animal nutrition research, and organizing resources for publications. The calorimeter team photographs always included the designated Institute of Animal Nutrition librarian of that year, and although their work was uncited in individual publications, they were listed as important staff in the Pennsylvania State College bulletins.[28]

EVERY COW A FACTORY

During the Progressive Era "efficiency" was used as a tool for social and political reform. Anxieties about wasted energy, money, and time came to the fore by 1910, which historians credit with the public attention gathered around Frederick Winslow Taylor's work on scientific management.[29] Just as motion studies became important ways to investigate wasted time and energy in the steel mill, so feeding experiments conducted by nutrition scientists helped investigate wasted food as it was processed in the body. The calorimeter was thus instrumental for not only providing more insight into the claims of feed manufacturers but for investigating better food production efficiency through certain kinds of animal feed and feeding.

Like other Progressive leaders of the time, Armsby was concerned with issues of food production efficiency in a world of limited resources. He voiced anxieties about population growth and food availability in several publications throughout his career.[30] His president's address for the 1909 American Society of Animal Nutrition meeting focused on what he called "late warnings of a deficiency in the food supply" for the future of the United States. Concerned about a population that could reach "twelve hundred millions" and an increased struggle for land with the closing of the frontier, Armsby pointed to animal nutrition science as a place to find answers to this looming challenge: "It would be foolish in the extreme to close our eyes to the fact that the intensity of the demand for food by our future population will exceed anything we have yet known. Whether this state of affairs is to come about more or less rapidly is important chiefly as it gives us more or less time to prepare for it . . . there are aspects of this question which intimately concern us as stock feeders."[31] What was the answer to ensuring food security for a growing national population? For Armsby it was animal feed efficiency. The work of animal scientists was the foundation for this answer, and just as Armsby firmly advocated for the legitimacy of Agricultural Experiment Station research during the proposal of the Hatch Act of 1887,[32] he pushed universities and the federal government to continue to fund feeding research for the sake of the nation's future.

To achieve feed efficiency, Armsby was specifically concerned with combating wasteful feeding practices. His interests can be broken down into three main strategies:

1. Promote foods for animals that are unsuitable for human consumption, that is, by-product feeding;

2. Promote research in animal physiology to understand the adequate metabolism of feed;

3. Fight against fraud in the feed industry.[33]

For Armsby testing feed company claims against fraud was but one way to obtain more efficient feeding. Researching how animals best converted feed for food production purposes was just as important for the efficiency mission. In his speeches Armsby emphasized that it was animal nutrition scientists, and the work of the American Society of Animal Nutrition, that would aid in finding the best by-product feeding practices.

The prospect that human food was being wasted on animals made for seductive national headlines. Armsby's speech took on a life of its own in newspapers across the country, with headlines reading "Fodder for Cattle, All Grain for Men" and "Animals Eat Our Food" published across Pennsylvania, the Midwest, and even California.[34] Anxieties about population numbers and the availability of bread were printed alongside Armsby's work and his suggestion to shift animal feed rations to by-product and crude fodder mixes. The calorimeter was showcased in these texts as the instrument that would help solve the problem of human food wasted on animals.

Armsby's calorimeter made efficiency measurable. But what exactly was being measured? The calorimeter measured two things simultaneously: bodies and feed *in* bodies. Each object needed to be investigated with thresholds and standards in mind. In making a feed ration, materials contributed to different bodily processes and manifested in three main ways. First, feed sustained the body of a given animal, allowing organs and bodily processes to operate seamlessly. Second, feed contributed to meat and milk production. Finally, there was feed excess, material not necessarily processed for sustenance or food production but expelled from the body as waste. Scientists were interested in gaining a foothold on this efficiency balance to create a feed ration that sustained life while maximizing production and decreasing waste. To do this it was essential to standardize what scientists referred to as "bodily maintenance."

At this time "maintenance" in agricultural research signified different ideas. "Maintenance research" entailed extensive work in biological tinkering: investigating how plants and animals were adaptable, able to withstand problematic conditions, and produce higher yields through human intervention.[35] Feeding research fell under the umbrella of maintenance work, but disciplinarily speaking "maintenance" meant (and continues to mean) something very different for animal nutrition scientists. Maintenance was not a type of research or a means to adapt. Maintenance was a bodily condition. It was a universal state across species and body types. It was a number. For cattle Armsby claimed maintenance was 6.2 therms or one-half pound of crude protein.[36]

Animal nutrition scientists' search for the maintenance ration was an exercise in rendering animals into abstract, standardized technologies.[37] When defining "maintenance" in his own reports, Armsby found it helpful to engage with a factory metaphor: "The animal mechanism must be provided with sufficient feed to maintain the processes essential to life before any continued production is possible. The amount of feed required for this purpose is called the maintenance ration of the particular animal. . . . To recur to the illustration of the factory, the maintenance ration keeps the empty machinery running, while the additional feed furnishes the power necessary to turn out the product."[38] Though the body-as-machine metaphor was popular for describing how food worked as "fuel," for both humans and other animals, since the nineteenth century,[39] it is striking that the animal body was more than mere machine in this description. Food animal bodies were full-fledged *factories* of food production.[40] Although full farm systems had not yet fully embraced the factory model at this time in history, as Deborah Fitzgerald has shown,[41] early twentieth-century scientists were actively thinking about individual animals as factories of food production.

Scientists also compared cattle to other forms of technology. Returning to Armsby's publications, he sometimes referred to feed as "solar energy" and cattle as trains harnessing this power from the sun. He cited the locomotive testing plants for the Pennsylvania Railroad in Altoona to provide this visual for readers of *Popular Science Monthly* to illustrate the purpose of the calorimeter.[42] Efficient food production by animals was thus compared to other goals of Progressive life, including the crackdown on railroads over wasteful spending and the push for efficient thermal and mechanical inputs and outputs.[43] The calorimeter recalled investigations for railroad efficiency standards, with Armsby's and others' maintenance ration calculations acting as standards to build more efficient feeding plans for food-producing animals.

Whether considering cattle as organic machines, food production factories, or solar-powered trains, Armsby's calorimeter measured living animal technology.[44] Reducing the animal body to numerical inputs and outputs, including its maintenance threshold, allowed scientists and farmers to reduce the complexities of metabolism into checks and balances. While the Pure Food and Drug Act promised to keep food pure for the sake of efficiency, studies with the calorimeter—in both humans and cattle—promised calculations for food valuations and rations that best fostered efficient bodily productivity.

HARNESSING KNOWLEDGE FROM "HAPPY BEASTS"

Visualizing animals as technologies and feed as fuel made the process of standardization for the sake of national feeding recommendations more manageable. However, scientists created these standards based on a set of choices regarding the individual animals they used for their experiments. As abstract and replicable as these animals and their behavior appeared in the USDA publications and *Farmers' Bulletins*, the test subjects both mirrored and contrasted with those animals found on the average early twentieth-century farm. Scientists relied on a very particular set of characteristics and forms of cooperation in their cattle, partially based on what they needed from animals participating in calorimeter experiments. While considered "happier" beasts than animals subject to vivisection, the newness of the calorimeter led to a series of trial-and-error interactions between experimental animals and the scientists who studied them.

The steer featured at the beginning of this chapter was the first recorded bovine to be tested inside the calorimeter's chamber. He was a three-year-old shorthorn steer, unnamed in the publications that mentioned him but presumably brought to State College from Pittsburgh like many of the farm animals for the experimental college.[45] Armsby and his team dedicated a significant amount of time training the steer to stand in a digestion stall before entering the calorimeter. The stall helped acclimate him to the size of the space he would have to stand, lie down, and eat within the device.

A testament to the shorthorn steer's value after this training, Armsby used him in feeding experiments from 1901 through 1904. Each experiment built on the last and grew more complex in execution. The Shorthorn steer even experienced a kind of luxury during his time as an experimental animal. To consider the growth of epidermal tissue in 1903 experiments, for example, the steer received thorough brushings before and after his stay in the calorimeter chamber.[46] Brushing not only allowed the scientists to collect his dandruff for further analysis, but also calmed the steer and reconstituted positive bonds between himself and the human scientists.[47]

However, not all test animals were as seasoned as this steer. Animals in the past, as they are today, were unpredictable, and some created challenges for the Penn State scientists in their mission to measure feed values and metabolic efficiency. Steer No. 522, for example, was not entirely satisfied with his feed ration during the Adams Act "Use of Food" experiment, which was completed as a joint effort between Penn State and the Missouri Agricultural Experiment Station in 1907. The experiment was meant to calculate the energy of Missouri's alfalfa feed, and during it No. 522 was fed the maintenance ration of this feed.[48] On October 13, 1907, three months into the experiment and before calorimeter calculations were made, he was found outside of his pen, to the dismay of the scientists: "No. 522 was found in the hallway this morning. Apparently he had eaten all the hay he cared for. Therefore he was not fed grain this morning. Someone probably went through the barn and left the gate to the hallway open, after I left the barn at night."[49] Many of these challenges outlined in the hand-written lab books never made it into the publications. Yet, there were some cases when the unpredictability of animals altered the scientific calculations and needed to be addressed in the formal write-ups. For example, Ox B from maintenance experiments between 1905 and 1907 had to be flagged as an outlier in a 1911 publication because of his lack of cooperation. Though he went through the training in the digestion stalls, he would not replicate his behavior inside the calorimeter. Ox B was so anxious, he would not lie down in the device, making his maintenance calculation well above those of his fellow oxen.[50] Such behaviors compromised experiments, but they also reflected the true unease felt by animals inside the small, confined space, even after proper training. The repeated use of the same animals in experiments represented the importance of such training and comfort, with the use of Ox B for multiple calorimeter runs likely illustrating a hope that the bovine would become more acclimated to the technology over time. This decision-making weighed the usefulness of the test subject, the significance of the test, and the wants and welfare of the animal—a complicated philosophical dilemma alongside a scientific one in these cases of clear animal anxiety inside the box.[51]

The Institute of Animal Nutrition completed its first grouping of calorimeter experiments primarily with male animals. Though their conclusions helped provide a foundation for understanding bovine ruminant metabolism in feed-to-food conversion, Armsby was much more interested in testing the metabolism of dairy cows. Scientists hypothesized that dairy cows were the most efficient food animals in converting raw materials into human foods.[52] Armsby cited this hypothesis extensively, and he wanted to further prove this case using his scientific device.[53] Though the team was able to complete a few trials with dry, pregnant animals, they realized too late that the body of the female bovine was not fully accounted for in the design and construction of the calorimeter.

Early media exposés of the calorimeter suggested that automatic milking machines would be used inside the device to provide more accurate measurements of milk production yields.[54] Whether this was proposed for the future by the scientists or solely a rumor made by journalists, milking machines were still being tested extensively at this time with inconclusive results.[55] Unlike the carefully crafted feed box, there was no airlock system designed into the calorimeter to account for the attachment and removal of such a device. If a cow was placed inside the calorimeter chamber, there needed to be a way to milk her without disrupting the artificial environment.

On June 1, 1916, the team investigated a workaround to their milking dilemma. The experiment was an attempt to distill the respiratory activity of a man from the cow he was milking inside the calorimeter chamber. Penn State technician Jons August Fries made a special respirator that he secured tightly onto the face of his test subject, Mr. Decker. Mr. Decker and Cow #579 were placed in the calorimeter together while Armsby and Fries went to work tinkering with the device's airflow meters to regulate the environment as the calorimeter door was opened and closed. They collected all possible fluctuations, including the outputs of Mr. Decker as he hand-milked Cow #579.[56] The experiment was haphazard but otherwise successful. Since reconstructing the calorimeter was, as Fries described, "out of the question,"[57] the scientists determined that the calculations could consider these twice-a-day outlier periods. They also concluded that fluctuations in the highly regulated air temperature were no great loss, with the changes well documented and accounted for in their metabolic calculations.

In addition to the problem of milking, the collection of cow urine and feces was more difficult than with steers because of the proximity of the urethra and anus in female cattle. As mentioned earlier, the steers participating in experiments wore special harnesses attached to their midsection which funneled their urine for easy collection by scientists. They also wore manure chutes behind them for the same purpose of excreta collection and analysis. These devices were worn by the animals in both the digestion stall and inside the calorimeter, allowing them to adapt to the straps and fabric and comfortably stand and lie down in each setting. Given cows evacuated urine and manure in the same relative area of their bodies, a new system needed to be put in place.

For experiments with cows Armsby and Fries tested various methods to collect excreta with hopes of finding one method that could be used for both the digestion stall and the respiration calorimeter. One idea they tested was a "dung box."[58] Fries created a multilayered container that would sit behind the cow and collect feces and urine together. A wide wire mesh basket sat on top of a box with a tightly bound layer of cheese cloth stretched in between them. This layer was meant to act as a sifter and hold any excess manure particles that bypassed the basket to allow only urine to filter to the bottommost section of the box. However, this idea and "various" others ultimately failed to adequately separate the materials.

The scientists ended up using two methods of excreta collection for their dairy cow experiments. First, in the digestion stalls, they tasked a watchman to manually collect the excreta with buckets as best as possible. Second, in the calorimeter, the researchers created a duct they attached to the backside of the cow so her excreta could be transferred to a receptacle outside the device. The duct was made in a way that allowed a cow to lie down within the device without it rubbing uncomfortably across her udder or milk veins. Although the duct was unable to adequately separate urine from feces, it was dubbed "very satisfactory" for the purpose of clearing cow excreta outside the calorimeter's regulated environment.[59]

INDUSTRIAL LOGIC FROM THE LAB TO THE FARM

In 1909 a farmer from northwestern Pennsylvania made the journey to visit Pennsylvania State College. He was curious to see what the campus was like and equally curious about the device that was repeatedly featured in newspapers across the state. He reportedly said that all the "time and trouble" to visit was worth it just to see Armsby's calorimeter.[60]

This interaction between a farmer and an exclusive scientific device was unique for this time in history. Agricultural "improvers," who conducted their own experiments and printed the results in self-published pamphlets, had dwindled in number.[61] Their authority had been replaced by the experiment station experts. Historians have also demonstrated that this period was a moment when the government questioned how likely it was for farmers to embrace the latest science and technology. High-profile responses to the "chaotic" inefficiency of farming in America included the Country Life Movement, which became an official commission in 1907.[62] The former's visit to the calorimeter proposes that farmers were more excited about these efficiency studies than government officials initially believed.

The documented visit by the farmer also suggests that the device reflected a new authoritative science for food producers among others across the state and the country. The calorimeter acted as a marker for the work and recommendations that made their way into the popular newspapers and farming magazines. Farmers were not the only ones who wanted to see the device in person. Feed manufacturers took advantage of opportunities to visit the college and see the calorimeter as participants of workshops and lectures about state feed analyses.[63] The workshops left time for participants to tour the device to better appreciate the authority of college lab results for given feed samples. German scientists eagerly visited in order to replicate Armsby's device in Bonn, Germany.[64] The World's Dairy Congress, hosted by the United States in 1923, even planned for international visitors to tour the college and see the calorimeter, a symbol of America's world renowned prowess in animal nutrition science.[65]

Though farmers and professional scientists alike visited the device out of curiosity, and perhaps because they were attracted to its novelty, the influence of its feeding conclusions are less clear. Did farmers readily adopt Armsby's recommendations and standards? How did Armsby's conclusions manifest as everyday practices on American farms?

At the legislative level, funding for experiment station work—including ensuring that public outlets for this new knowledge reached everyday farmers—was important to Armsby in his scientific mission. Historians consider his work instrumental not only for regulatory support of the 1887 Hatch Act but the later passage of the Adams Act, which increased Hatch Act funds to contribute to the extension work scientists completed with and for US farmers. Armsby led a letter-writing campaign in Pennsylvania to encourage representatives to pass the bill, and with the support of Pennsylvania's united farmers' organizations and the Pennsylvania Grange, the campaign was enough to annoy representatives.[66] With some credit to such letter-writing campaigns, the Adams Act was passed the same year as the Pure Food and Drug Act.

Complementary to this extension work, *Pennsylvania State Bulletins*, published by its agricultural station, including Armsby's publications, were featured prominently in widely circulated agricultural magazines like *Hoard's Dairyman*. In the editorial columns of this publication, farmers across the country asked for advice about balancing rations for their dairy herds. *Hoard's* editors often replied with "Armsby's standard."[67] This standard was attractive for Progressive agriculturalists because Armsby's calculations cited a lower net energy allowance for efficient food production numbers than other scientists of the time. Ideally, this meant less feed was required for bodily maintenance in calculated rations, with more feed contributing to meat or milk yields in cattle.

Armsby's experiments also provided evidence for long-standing theories farmers had debated regarding connections between animal care and production. For instance, it was unclear if steers needed to be housed in heat-regulated stables during cooler weather. Newspapers interpreted Armsby's initial respiration calorimeter reports in different ways on this issue, most claiming that Armsby's work showed that a steer's body heat was sufficient and that farmers did not have to invest in specially warmed stalls.[68] The *Lima News*, circulated in Ohio, confidently wrote as it cited the respiration calorimeter, "It may be said, therefore, with the authority of science and experience both, that the open lot with shed for shelter is the best place to fatten a steer."[69]

The case for comfortable shelter was also emphasized in papers using the same studies. Armsby's studies made it clear that animals who lay down used less energy, allowing more feed to be dedicated to food production efforts. The *New York Times* ran a special plug that headlined, "Dr. Armsby Says They Lose Weight When Standing Up," pressing the need for farmers to ensure comfortable quarters that invited cattle to lie down. The coverage also concluded that respiration calorimeter experiments provided "confirmation of the theory that quiet and contented animals make best gains."[70] How energy was used as animals lay down and stood up suggested that temperament also mattered in food production.

The scientific reports by Armsby and his team thus suggested clear connections between feed efficiency and "good breeding." Uneasy, anxious animals used more energy than calm and docile ones, and this was energy wasted on temperament rather than used to produce human food. Armsby noted in one report that restlessness could be caused by external circumstances. He cited a respiration experiment conducted on a horse in Germany, where flies caught in the respiration chamber caused an increase in the amount of material oxidized by the animal.[71] However, temperament was also a by-product of breeding. In his report on the calculation of the maintenance ration in cattle, Armsby addressed the problem with Ox B—the experimental animal who refused to lie down in the calorimeter—with reference to breeding. Armsby explained in his conclusions that Ox B was "a scrub of decidedly nervous temperament" and a sharp contrast to the model measurements produced by Ox A: a pure-beef animal who demonstrated the ideal "quieter and more phlegmatic" disposition.[72]

Armsby's calorimeter-based conclusions—encouraging meticulously developed feed rations, types of animal housing, and breed types—set a foundation for the later industrialization of US food animal agriculture. Much of what the Institute of Animal Nutrition proposed favored larger-scale operations focused on commodity production using new technologies, outsourced products, and outside expertise. At that time such recommendations also favored elite agriculturalists. For example, the purebred animals favored in the calorimeter experiments were primarily owned by upper- and middle-class farmers. "Gentleman farms," including Henry Francis du Pont's Winterthur of this time, housed many of the prize-winning herds in the country.[73] The wider availability of purebred animals would come only later with the development of new breeding technologies, including frozen semen that allowed type-breed cattle to become available across geographic boundaries that once restricted farmers' access to certain animals.[74] Later reliance on breeder services by industrial-level operations can be ascribed, at least in part, to these earlier scientific recommendations favoring purebreds.

Through the first half of the twentieth century various institutions, in addition to newspapers, agricultural journals, and experiment station bulletins, helped popularize scientific findings and the new agricultural practices attached to them. Agricultural exhibitions and 4-H programs reached the average farmer through fairs and youth groups to introduce different farm management practices to the average farmer. As knowledge circulated through these special outlets, both established and up-and-coming businesses took advantage of the new information to help sell services and products to farmers.[75]

In dairying, milking machinery companies like De Laval used their monthly publications to highlight the importance of feeding cattle for desired milk yields. Citing USDA *Department Bulletins*, magazine spreads encouraged farmers to "Listen In on the Herd," and invest in supplemental feed, feed manufacturing technologies, and De Laval milk scales to check feeding progress with subsequent milk yields.[76] These elaborately illustrated magazine spreads positioned De Laval products as just as essential as feeding and even breeding practices in dairying.

In efforts to elevate their legitimacy and secure long-term consumer relationships with farmers, animal feed companies also took advantage of the latest science and even reconfigured their business operations based on nutritional research. By the 1920s leaders from companies like Purina Mills encouraged feed manufacturers and salesmen to take on an advisory role in farming communities. They were to embrace the role of nutritionists and convince farmers to purchase their products based on recommendations from places like Penn State.[77] Feed salesmen embraced nutrition language in their regular one-on-one meetings with farmers and integrated it into new kinds of advertising literature. Advising booklets published by feed companies from as early as 1920, for example, featured Armsby's ration tables and included renderings of cattle as factories. Purina's *A-B-C of Milk Making* booklet even transposed an image of a cow on top of a factory with accompanying sections that included "A Cow Is a Milk

Factory" and "To Heat and Maintain a Cow."[78] Through this business literature and accompanying advisory services, feed companies used calorimeter results to solidify their place in everyday farming operations. This led farmers to increasingly rely on a network of experts and companies to feed their animals, a stark change from prior feeding arrangements that relied more heavily on the resources produced on the farm.

SHARING SPACE

The Institute of Animal Nutrition calorimeter experiments forced scientists to interact with cattle in ways that had not existed before. Feeding experiments relied on a technological instrument that required new, obsessive attention to the body, from feed inputs to excrement. In many ways the calorimeter created closer physical and intellectual proximities between humans and their domesticated farm animals. The fact that the same calorimeter used for cattle experiments was later used for human trials at Penn State speaks to this closeness and how shared concerns about nutritional values and outputs oscillated between animal and human experimentation.[79]

However, the conclusions that were made about feeding cattle using the calorimeter, regardless of the messy ways they were reached, reinforced the industrial logic that laid the foundation to industrialized cattle production as it is understood and practiced in the United States today. Farmers continue to meticulously measure feed inputs in efforts to manipulate their meat and milk yields. The menus they feed their animals rely on scientifically informed feed manufacturing practices and require supplemental products sold by feed companies. Arguably, industrialized agriculture has created more gaps than shared spaces between humans and their food animals, particularly for the human consumers of the products animals create. Yet looking closer at how humans and cattle interact on farms, and how this has changed through a longer history that includes the use of the Armsby calorimeter, complicates what closeness, care, and cooperation has meant between humans and their food animals. The calorimeter enabled bridges to be built between humans and animals in nutritional research and fostered on-the-farm efforts that paid closer attention to how animals were fed, housed, and bred. All the while, calorimeter research accelerated values in industrial agriculture that we need to continue to reevaluate, as its outputs have created more ecological challenges than imagined during Armsby's time.

Calorimeters are still used today but with increased attention to metabolic methane outputs.[80] This focus on respiratory outputs in conjunction with beef and milk production speaks to the importance of studying technologies and animals within their context. Although it arguably aided in the creation of the problems that led to the excess bovine enteric methane studied today, the calorimeter presented in this chapter was ultimately a technological site where humans and bovines came together to explore problems. These instruments continue to be the sites of shared experience, insight, and exploration, and continue to require docile, domesticated cattle compatriots to complete the scientific work. The long-standing challenges in connecting the individual and the collective, the autonomous to the standardizable, continue to plague scientists as they attempt to translate individualized data into national and international recommendations. As with the story of Armsby's calorimeter, technological instruments may demonstrate our shared stake in more sustainable and humane food futures. However, if not careful, the conclusions this research creates may continue to widen the gap between human interests and the lived realities of their animal compatriots.

CHAPTER 2

PROTOCOLIZING HUMAN-ANIMAL RELATIONSHIPS IN COLONIAL QINGDAO

TATSUYA MITSUDA

During the late nineteenth-century scramble for colonies, Imperial Germany gained control of Jiaozhou Bay (German: Kiautschou) and developed the leased territory on the Shandong Peninsula in eastern China into a strategic naval base in East Asia. On their arrival in 1898, the Germans also set about transforming the small fishing village of Qingdao (German: Tsingtau) into a model colonial city to rival the British possession of Hong Kong.[1] As economic historians have shown, Qingdao, which boasted an ice-free port, quickly became embedded in world markets, functioning as a commercial hub that exploited the natural resources of northeastern China for international trade.[2] What is less well known about the colony is that imperial expansion and economic activities included not only natural resources such as coal, cotton, and silk, but also extended to the exploitation of animal resources such as cattle. Responding to the need to ensure a safe, stable, and sufficient supply of milk and meat, first the Germans and then the Japanese (who took over the colony after the outbreak of World War I) constructed and reinforced scientific and technological management systems that monitored, controlled, and disciplined the interaction between human and animal bodies to prevent disease outbreaks. Racial prejudice informed the creation of these systems, as colonizers sought to correct the free, haphazard, and dangerous relationships that were seen to exist between the Chinese and their animals. Failure to bring them under control risked producing dirty, contaminated, and disease-ridden milk and meat. Through an analysis of the veterinary inspection regimes implemented in Qingdao, this chapter reveals how the Germans and the Japanese responded to this "problem" by reorganizing the spaces in which human and nonhuman animal bodies interacted, thus creating a more disentangled, controlled, and disciplined relationship—a process this chapter will characterize as "protocolization."

In their introduction to this book, Finn Arne Jørgensen and Dolly Jørgensen point out that technological mediation has been key to making individual animals visible in recent times. They argue that technological management systems such as GPS have made it easier to tag, monitor, and manage the movements of nonhuman animals, allowing the broader public to gain access to the life and death of individual creatures such as Bear 71 despite the barriers of time and space that previously prevented the forging of such relationships. In the nineteenth century, by contrast, the scientific and technological regimes that were put in place to process food animals, even as their numbers increased in urban areas, tended to have the opposite effect—to make them increasingly invisible.[3] Nowhere was this more evident than in the emergence of modern slaughterhouses. As historians have shown, a combination of middle-class squeamishness about the sight of death, distrust of the butchers who performed slaughter and inspection, and the logistics of feeding an increasingly urban population conspired to reduce the number of private slaughterhouses and push them out of town and city centers.[4] Long-distance railroad networks also contributed to this disappearance, as livestock that had previously been walked to markets were shipped as freight to large abattoirs, where, under the auspices of public authorities, scientifically trained experts increasingly performed the task of inspecting and processing animal bodies—both alive and dead—away from public view.[5]

Focusing on the Chinese cattle destined for German and Japanese exploitation, this chapter examines how colonial bovine bodies became embedded in scientific and technological management systems that reflected shifting imperial interests and commercial rivalries in the region. It shows how and why animal bodies were increasingly subjected to surveillance and control through registration, inspection, quarantine, and inoculation designed to control the spread of infectious diseases. More specifically, it argues that as cattle became more important for human use, minute details came to govern the interaction between human and animal bodies, resulting in an intensified and highly protocolized human-animal relationship that was unique in East Asia at the time. Since then, as in other parts of the world, such scientific and technological management systems have become the norm, which wider society sees only when similar controls are placed on human movements.

DISENTANGLING CHINESE HUMAN-ANIMAL RELATIONSHIPS

When the Germans arrived in Qingdao, the settlers faced the immediate challenge of securing a food supply that would allow them to maintain their European diet. Before their arrival, the Prussian geographer, Ferdinand von Richthofen, who is widely credited with recommending this part of China for imperial exploitation, had told his compatriots that eating like the natives was not an option. As George Steinmetz has shown, Richthofen, a frequent traveler to China in the late 1860s and early 1870s, had an unflattering view of the locals he encountered, describing them as "slanty-eyed angels" who emitted a "specific odor that is unique to the race."[6] A European would never develop true bonds with the Chinese, he sniffed, "except in the form of the relationship between a master and his dog."[7] Such a racialized view informed his assessment of the native diet, which Richthofen could not bring himself to taste, describing it as "unpleasant and unbearable."[8] Referring specifically to Shandong Province, of which Qingdao was a part, Richthofen wrote that Europeans would largely have to secure for themselves the ingredients for Western dishes. But access to the right supplies was difficult, especially when it came to sourcing animal-derived ingredients. While beef, chicken, and eggs could be purchased at markets, pheasants, quail, and wild doves had to be hunted. However, he sternly warned his readers to avoid pork. Unlike other meats that were not as popular, the fact that pork was readily consumed by the natives made him suspicious of its "unclean nature."[9] An even greater problem was the lack of milk. Since the Chinese were not in the habit of drinking dairy milk, the geographer had to make do with the "condensed milk" he had brought with him, which he poured into black tea instead of green tea, which he avoided drinking.[10] What Richthofen ate and drank was thus a conscious attempt to maintain his European identity and racial hierarchy—an attitude that his compatriots adopted when they arrived in Kiautschou.

Of the two supply issues, the German colonizers prioritized access to fresh milk. Upon arrival in the Far East, German mothers struggled with acclimatization. They had difficulty breastfeeding their babies due to living in a different environment and climate, creating an urgent need for dairy milk that could be quickly prepared as either a supplement or replacement.[11] Sourcing milk locally, however, meant that the Germans had to rely on the Chinese, who were suspected of either hiding the true health of their dairy cattle or diluting the resulting milk with lime water and selling it to the Germans at exorbitant prices. To put an end to the adulteration, the Germans implemented regulations in 1899 that placed the interaction between the indigenous population and their cows under stricter control. Cows destined for German consumption had to be registered, inspected, and deemed fit for use; failure to remove or report suspicious cases resulted in harsh penalties.[12] Regulating the milk supply, however, did little to solve the problem of volume. Unlike European breeds, which had been engineered through selective breeding to optimize production, Chinese breeds had not undergone similar biological intervention and could produce only between 1.5 and 2 liters of milk per day because of shorter and slower lactation cycles. To speed up the process, the colonizers thus decided to import East Frisian Jeverland bulls and assigned veterinarians to crossbreed them with Chinese cows in the hope of increasing the productivity of the resulting offspring. Such management systems, which turned scientific experts into surveillance agents, allowed the Germans to take the first steps toward exercising greater control over colonial bovine bodies, including intervention in the reproductive process itself. Compared to milk, the procurement of meat, especially beef, did not require such drastic intervention: instead of importing German or other Western cattle, abundant supplies could be procured from the "hinterland" of Shandong Province for processing into meat. However, reflecting racial fears about edibility and hygiene, the Germans moved to police the trade, with veterinarians tasked with establishing a meat inspection regime (*Fleischbeschau*). Ernst Rassau, the colony's veterinary officer, was initially skeptical about implementing such a sophisticated management system that had been developed for a more modern, populous, and civilized society. But the "disgusting" way the locals interacted with the animals quickly changed his mind. He vividly recalled the sight of Chinese butchers "gleefully consuming an abscess" from a diseased animal "to show me that it was safe to eat."[13]

Initially spatial and temporal controls were rudimentary: inspections consisted only of periodic visits by the veterinarian to the surrounding villages where slaughter took place.[14] Visits were vaguely scheduled in the mornings and afternoons, and regulations were inconsistently applied. Emergency slaughter—the practice of slaughtering injured or sick animals almost anywhere and at any time—remained common. Under this regime, post-slaughter inspection was narrowly focused on the carcass, with the scientific gaze looking at "the organs, in particular, the lungs, liver, and kidneys" as well as the udder and uterus of cattle, to determine whether the carcass was fit for consumption.[15] There was no preslaughter inspection, no monitoring of the health of cattle in stockyards, no lengthy quarantine, and no inoculations. Most important, a shortage of scientific personnel made it virtually impossible to monitor and control the movement of animal bodies throughout the colony and beyond, which is why Rassau called for the construction of a modern slaughterhouse as the centerpiece of the meat inspection regime. "Due to insufficient numbers of personnel tasked with surveillance and monitoring," he wrote, "comprehensive controls" would be possible only if, after the construction of the slaughterhouse and stockyard, "all suppliers were required to have slaughter done there."[16]

When the slaughterhouse began operations in 1905, the spatial and temporal controls that governed human-animal interaction were tightened. At the center of the slaughterhouse grounds were the slaughter halls (*Schlachthallen*), which measured 76.34 meters in length and 42.60 meters in width.[17] Each of the two main halls—one built specifically for large animals such as cattle, the other for small animals such as pigs—provided a highly sanitary space for processing livestock. A phalanx of separate rooms, seamlessly connected to the slaughter halls, further enhanced the efficient and hygienic processing of animal carcasses. These included two tripe cleaning rooms, two dung houses, a room for slaughterers, a bathroom, a workshop, an engine room for making ice, a steam boiler room, a cold room, a trichinosis inspection room, an equipment room, and the director's office.[18] Slaughterhouse architects paid particular attention to the circulation of air and water, which could become carriers of pathogens. The 1.5-meter-thick walls were therefore built with two layers of insulation added to minimize the expulsion of potentially contaminated air. To improve ventilation, the interior was also equipped with thirty-three suction devices consisting of valves and shutters that allowed outside air to enter.[19] Water flow was also tightly controlled to reduce the risk of infection. Separation of the municipal water supply from the seawater supply was of paramount importance, with the latter being used primarily to flush out waste that accumulated in the slaughter halls.

PROTOCOLIZING HUMAN-ANIMAL RELATIONSHIPS

Such spatial and technological infrastructures helped to protocolize human-animal relations. Arriving cattle had to be inspected by a veterinarian before entering the stockyard, where they were marked "fit for slaughter" (*schlachtbar*).[20] Those deemed "suspect" were quickly separated from the rest of the herd and taken to a separate slaughterhouse for sick and diseased animals, where they remained until their entrails were deemed fit for consumption. Timing rules also determined the cutoff point for inspection: cattle arriving for slaughter in the morning had to have passed veterinary inspection before 6 p.m. the day before; those scheduled for slaughter in the afternoon, on the other hand, had to be presented for inspection before 10 a.m. on the day of slaughter.[21] When it was time for cattle to enter the slaughterhouse compound, the movement of cattle was restricted to two specific times of day: from April 1 to September 30 cattle could enter only between 6 and 10 a.m., and then between 4 and 6 p.m. Between October 1 and March 31 the window shifted to a later time of 8 to 10 a.m.—a reflection of the need for sunlight to guide cattle into the compound. Slaughter times were aligned with the entry times, with slaughter running from 5 to 10 a.m. and 3 to 6 p.m. in the summer and from 7 to 12 p.m. and 3 to 6 p.m. in the winter. The duration of slaughter was also fixed: it took exactly one and a half hours to slaughter, dismember, and inspect the resulting body parts.

Strict protocols also governed the movement of people and vehicles entering the site. To minimize the risk of infection, the rules stipulated that only those directly involved in the slaughter process, including companions, could enter; they had to apply for a permit and always carry certification. Once inside the slaughter halls, companions were allowed only a short stay; they were told where to sit and were required to follow instructions detailing what they could and could not do.[22] The rules governing human behavior extended beyond the slaughter. Only after the carcass was completely dried and cooled could companions take the meat to refrigeration facilities. The cutting of the meat itself was allowed only in the refrigeration rooms, and the knives and swords used for this purpose were to remain there. When it came time to transport the meat, it could be transported only in closed cars or covered with clean cloths. Even the vehicles used to transport the equipment had to follow specific rules about where they could be placed, how long they could stay, and the speed at which they could be driven.[23] Such rules had the added benefit of helping to discipline the "disobedient" Chinese. Eggebrecht was pleased to report that the behavior of the Chinese slaughterers could be monitored, controlled, and limited to practices that conformed to the sanitary protocols established by the Germans.[24]

Although the meat inspection regime centered on the slaughterhouse was designed primarily to protect the health of the colonists, the opening of the German-built railroad linking Jinan to Qingdao internationalized the meat trade and changed the context in which cattle slaughter took place. Completed in 1904, the railroad allowed cattle to be transported directly from the main livestock market in Jinan, making it faster, cheaper, and more standardized. Before its completion the journey was estimated to take ten to twelve days, but afterward this was reduced to just twelve hours.[25] With Jinan also serving as the main collection point for livestock from throughout Shandong Province and beyond, the potential market for livestock trading was immense, a development further accelerated by the opening of new railroads, such as the one between Jinan and Pukou, near Nanjing. Initially the number of animals being shipped was small, although the Shantung Railway Company (Die Schantung-Eisenbahn-Gesellschaft) had prepared twenty-five livestock cars in anticipation of demand. One reason for the lack of interest, the company wrote, was the habit of transporting livestock by road and water, a practice that the company was determined to change.[26] When the slaughterhouse began operations in 1905, freight volumes increased rapidly. A year later the number of animals transported by rail had quadrupled, prompting the company to remark that livestock freight was becoming "a stable source of income."[27] In 1908 the number of large animals shipped by rail reached a record 19,428. It should be noted, however, that the cattle (which made up the bulk of the large animals) were not for domestic consumption but largely for export. In 1906 the Russians quickly realized the potential of livestock transportation, shipping 2,081 head of cattle and 565 head of sheep from Qingdao to Vladivostok.[28] A few years later the Russians began exporting meat, which led to the expansion of refrigeration facilities at the slaughterhouse. In 1914 more than 45,000 animals were slaughtered in Qingdao, most of which were exported.[29]

For the more locally minded German colonists charged with protecting animal health, however, this boom in livestock shipments from the interior of Shandong Province was more cause for angst rather than celebration. Because the arrival of increasing numbers of Shandong cattle posed a sanitary threat to

Qingdao, especially to the breeding and maintenance of dairy cattle, veterinarians feared increased exposure to endemic infectious diseases such as rinderpest, which was endemic in northern China. As mentioned earlier, the colony's milk supply depended on the importation of Western breeds, which were crossed with Chinese cattle to increase yields and keep bovine bodies firmly under German control. Repeated attempts to import Jeverland cattle from Germany did not always go according to plan, however, as few sanitary precautions were taken to protect the animals. After the railroad and abattoir were opened, German and Australian breeding stock died of rinderpest.[30] A few years later, two bulls and three cows were imported, but contracted the disease on the way to Qingdao.[31] A better management system was needed to control the movement of Shandong cattle and to monitor the spaces where pathogens could enter.

In 1909 Max Eggebrecht, Rassau's successor, called for comprehensive controls on the movement of cattle to avert the threat of infection to German dairy herds.[32] He required railroads to transport all cattle destined for slaughter, proposed designated areas for stockyards, ordered border closures of surrounding areas, established new quarantine rules, and called for the registration and inoculation of all cattle in Qingdao. "To carry out the breeding experiments successfully," he wrote, "it is absolutely necessary to erect barriers to sanitize parts of the leased territory."[33] In making these suggestions, Eggebrecht emphasized the need to move cattle off the roads and onto railroads, subjecting their bodies to constant surveillance. Noting that land routes were still being used for cattle coming from neighboring areas such as Tschautsung, he warned that the result of this "unhindered free movement of cattle" increased the risk of epizootic outbreaks in the borderlands.[34] Cattle arriving by rail from the interior of Shandong, the veterinarian further recommended, should be quarantined near the small port station at Ritthausen, which was considered ideal because of its geographical advantages: it was flanked on the north, south, and west by the sea, which provided a natural defense. Police stations in the small port at the main railroad station would also make the task of surveillance and monitoring easier. Despite his best efforts at persuasion, Eggebrecht's proposals met with stiff resistance from the medical officers, who felt that the veterinarian's proposals went too far in extending sanitary controls beyond the city limits. To them, inoculating cattle, for example, was something that should be limited to Western cattle arriving in Qingdao and not offered to Chinese cattle at all.[35] Nevertheless, German veterinarians continued to research and develop sera, which were then made available to the Russians.[36]

Much of this reluctance to expand controls was rooted in the colonizers' low regard for livestock as an economic resource to be exploited for international trade. Compared to its colonies in Africa, the Chinese colony was largely a commercial, not an agricultural, enterprise. Local peasants were mostly subsistence farmers who were not interested in breeding and raising animals for larger markets, resulting in a very limited number of cattle, mules, and donkeys that were used primarily as draft animals. In Eggebrecht's view the absence of large landowners—a common presence in Europe—meant that the kind of leadership that had helped develop the livestock industry in the West was simply lacking in China.[37] While the German-Chinese University (Deutsch-Chinesische Hochschule), which opened in 1909, aimed to improve Chinese agricultural production, its establishment came too late to make a significant difference before the end of German colonial rule. Nevertheless, the Germans had succeeded in establishing a scientific and technological management system that closely monitored the movement of colonial cattle destined for local use. Most important, inspection regimes for milk and meat replaced Chinese middlemen with German scientific experts, who henceforth reserved the right to decide the fate and value of livestock. The changes ushered in by the opening of a modern slaughterhouse were marked by the creation of new spaces in which incoming Chinese bovine bodies would be subjected to unprecedented and intense scrutiny, with detailed rules governing where, when, and how they would be processed. In this way, a highly protocolized form of human-animal relationship emerged, a process that the arriving Japanese both intensified and expanded.

INTENSIFYING PROTOCOLIZED HUMAN-ANIMAL RELATIONSHIPS

When the Japanese took advantage of the outbreak of World War I to wrest control of Kiautschou Bay away from the Germans, they were less concerned with securing food supplies for the colony than with exporting cattle and beef to meet growing demand back in Japan. Since the Meiji Restoration (1868), when Japan began its modernization drive, beef had been elevated as a symbol of civilization whose consumption helped reform the largely plant-based indigenous diet to create strong Japanese bodies that could compete with Western bodies.[38] Despite concerted government-led efforts to improve, enlarge, and multiply the domestic stock through crossbreeding with Western strains, Japan struggled to make beef readily available for mass consumption.[39] Successive wars, such as the Sino-Japanese War (1894–1895) and the Russo-Japanese War (1904–1905), did not help; they periodically drove up beef prices as large numbers of domestic cattle were hastily slaughtered to supply soldiers with beef rations. Not surprisingly, the military arriving in Qingdao quickly became interested in Shandong cattle, an enthusiasm that extended to their breeding. Although the military authorities considered Chinese animal husbandry techniques to be rather "primitive," they praised the region's three-thousand-year history of animal husbandry.[40] Livestock breeding was widespread throughout Shandong, with the eastern regions considered to have strong traditions. And they compiled statistics to show that the province had a total of six hundred thousand cattle awaiting exploitation.[41]

In contrast to German indifference to the international beef and cattle trade, the Japanese understood that they were in fierce competition with other imperial powers for their share of the expanding business. They pointed not only to the Russians as the most active exporters of beef and cattle, but also to the British, who through the trading company Jardine Matheson were also busy tapping into the bovine resources of Shandong, transporting them by rail to Nanjing, where they were slaughtered and packed into refrigerated containers for shipment to the metropole.[42] Even the Americans, it was reported, were muscling in on the trade, taking Qingdao beef to the Philippines and other Southeast Asian countries, especially after beef supplies from Australia dried up with the outbreak of World War I.[43] Reflecting this enthusiasm, the military administration experimented by shipping nine hundred live cattle to the port of Yokohama in 1916. The Ministry of the Interior also sent its chief technocrat, Yamawaki Keigo, to Qingdao to discuss export plans with the army veterinarian Matsuo Hiroshi.[44] Statistical information was collected, feasibility studies were carried out, and a company called the Imperial Livestock Trade Company (Teikoku Chikusan Bōeki Kabushiki Gaisha) was established to prepare for the impending trade.[45]

Like the Germans, the Japanese colonizers expressed racial suspicions about how locals interacted with animal bodies and reaffirmed the need for scientific and technological management systems that would monitor, control, and discipline the Chinese relationship with livestock. As Yamawaki put it, the Chinese could hardly be trusted because they had "little scruples" about secretly slaughtering diseased cattle and offering them for sale.[46] Despite frequent epizootic flare-ups, the veterinarian continued, the Chinese were unaware of the health risks these outbreaks posed. As a result Japanese veterinarians welcomed the meat inspection regime, centered around the slaughterhouse, that the Germans had left behind, an appreciation made easier because they had been avid students of it for over twenty years. Paralleling developments in human health, where a German model of public health had been adopted since the 1860s, the Japanese also sought to implement a German model of animal health. With the establishment of a meat inspection regime centered on public slaughterhouses, Japanese veterinarians made regular visits to Germany to learn from their regulations and the slaughterhouses being built there.[47] Countless books and articles, some of which were translated into Japanese, had long been required reading for students and practitioners, who quickly warmed to what they saw in Qingdao. Marveling at the architectural scale and technical sophistication of the slaughterhouse compound, veterinarians like Matsuo wrote in glowing terms about how everything worked like clockwork to properly discipline the interaction between human and animal bodies.[48]

Given their praise for the German management system, it is not surprising that the Japanese left intact the protocols implemented by the Germans, choosing to make only minor adjustments, most of which focused not on the interactions within the compound that the Germans had focused on regulating, but on the "proper" relationship that should exist at the boundary between the slaughterhouse and the stockyard.[49] For example, the revised rules stipulated that companions should "gently stroke" (*aibu wo kuwaeru*) the animals before entering the compound, to avoid the commotion that typically accompanied cattle movement. Adjustments were also made to the route taken by animals on their final journey to the various slaughter halls.[50] Previously blood and feces had accumulated inside the facility, making the spaces between the slaughter halls and the stockyard unhygienic. To reduce the amount of dirt that could clog up those spaces, cattle were brought in through the side instead of the main entrance. The extra monitoring also extended to the stockyard itself. Under the Germans, preslaughter inspection had taken place outdoors, making inspection difficult and measurements inaccurate as bovine bodies were exposed to the vagaries of the weather. To solve this problem, a shed was built on the right side of the compound, where cattle were brought for a more efficient and reliable inspection.

INTERVENING IN COLONIAL BOVINE BODIES

Despite changes in German protocols, the Russians remained the main users of the slaughterhouse and exporters of livestock whose health was inspected by their own veterinarians. A combination of factors was responsible for this situation. First, refrigeration technology, while already in use by the British, Americans, and Russians, had not yet been adopted by the Japanese. There had been past attempts to encourage uptake, most notably when importers attempted to bring Australian beef to market in 1908.[51] But skepticism about "cold" meat, as opposed to freshly slaughtered meat that had been prepared near the place of consumption, worried traders that Qingdao beef might be similarly rejected.[52] Second, historic reservations about the threat of foreign epizootics held back mass exports to the metropole. A year after the Japanese took control, 34 head of cattle were found to be infected with rinderpest; the number rose to 84 in 1916 and 107 a year later.[53] Attempts to inoculate live cattle with Russian-made sera failed to achieve the desired level of effectiveness. At a meeting at the Ministry of the Interior in 1919, however, technocrats decided to respond to spiraling domestic beef prices by allowing imports on the condition that Chinese cattle be inoculated with a Japanese-made serum. Confidence in an expanding imperial veterinary network was key to the decision: the creation of a more effective serum, developed by Japanese veterinarians at the Rinderpest Serum Manufacturing Institute in colonial Korea, provided the technological breakthrough. An initial order of one million vials was placed in 1920, but with the establishment of the Shandong Drugs Research Institute (Santō Yakubutsu Kenkyūjo) in 1921, Qingdao became self-sufficient in serum production. Although the institute was a private enterprise subsidized by the government, it began production of rinderpest sera in 1921 and later expanded research into other diseases such as foot-and-mouth and swine fever.

Unlike the Germans, whose local concerns limited the scope of controls, the Japanese use of sera allowed for a broader and more interventionist range of controls. The export of live Shandong cattle also required the establishment of a management system that was as protocolized as the meat inspection regime already in place. Chinese bovine bodies destined for Japan were quarantined for a maximum of twelve days and subjected to numerous health checks and inoculations.[54] As in the slaughterhouse, the quarantine zones became places where cattle were monitored and the interaction between humans and livestock was strictly regulated. No one but the inspector was allowed into these areas, and the scope of inspections extended to checks throughout the cattle's stay.[55] Before entering the quarantine zones, cattle had to be inspected for suspicious symptoms. Once inside, they were further inspected, thoroughly disinfected, and placed in designated stalls where they awaited more intensive checks to certify their health for export. Throughout the quarantine period, the health of the cattle was constantly monitored. At 12 and 3 p.m. each day, their temperatures were taken, and the results of clinical examinations were recorded. Permissible body temperatures were set at 39°C, with exceptions made when environmental conditions and the individual's constitution were thought to have an effect. The timing of injections was also standardized: only cattle that had not shown symptoms for three consecutive days could be inoculated. The amount of serum administered was also carefully calibrated. For cattle weighing 300 kg, 150 mL was injected, a dosage that increased at 50 kg intervals. For a period of time after inoculation, cattle had to be immobilized, rested on the spot, and observed for any significant side effects. Finally, cattle that made it through quarantine, inspection, and inoculation would, on leaving the area for the stockyard, have any dirt that may have accumulated on their hooves removed, their legs doused with carbolic acid from the knee down, and their tails and anuses thoroughly disinfected. Similar to slaughter, living Chinese bovine bodies were subjected to an intense level of monitoring and surveillance before they could be passed as fit for export, ushering in yet another layer of controls that governed the relationship between humans and nonhuman animals in Qingdao.

A major problem with transporting livestock, as opposed to carcasses, was cost. Quarantining Shandong cattle for twelve days, conducting a series of tests, inoculating them with sera, and disinfecting their bodies cost time, labor, and money. Similar quarantine and checks, though not as rigorous, were required for cattle arriving in Japan before they could be finally released into metropolitan markets. Live cattle also weighed more, took up more space than refrigerated carcasses, and required feeding to minimize the loss of meat that was an unavoidable by-product of crossing the East China Sea. In addition, transporting livestock was riskier than shipping meat. A not inconsiderable proportion of cattle invariably died en route; hostile environments such as typhoons and sweltering heat led to these deaths. In 1923, for example, the ship *Tōgōmaru*, sailing from Qingdao to Yokohama, sank near Korea with 345 head of cattle on board.[56] Only a month later, a different ship, the *Kantonmaru*, carrying a cargo of 471 live cattle, arrived in Japan with 110 fatalities, a result of a combination of strong winds and scorching heat.[57] A further blow to this trade came when export duties were reimposed after the return of the leased territory to the Chinese, and restrictions were placed on the export of bulls to prevent the flight of animal labor from Shandong farms.[58] Coupled with the intense competition Qingdao faced from ports such as Tianjin and Dalian, Japanese merchants desperately petitioned the Foreign Ministry for action in 1924, but to no avail. After peaking at around 13,000 head of live cattle in 1920, exports of Shandong cattle via Qingdao plummeted to a record low of 307 in 1926.[59] In contrast, the amount of refrigerated beef increased rapidly after 1919, climbing to more than 60,000 head in 1923 and maintaining steady annual exports of more than 50,000 thereafter.[60] These cattle were invariably transported on the railroad that linked Jinan to Qingdao, arrived at the stockyard on the day before slaughter, underwent veterinary inspection, and on the day of slaughter underwent a highly protocolized process that transformed them into chilled meat ready for shipment.

Colonial expansion brought about a major change in the management of food animals in East Asia. Driven by the need to supply their populations with animal-derived foods, first the German and then the Japanese colonizers established a series of scientific and technological management systems that intervened heavily in the human-animal interactions, resulting in highly protocolized human-animal relationships that were designed to protect the health of the imperialists. The variety of controls and surveillance activities that these systems instituted reflected the colonizers' own set of priorities. Locally oriented, the Germans focused on the immediate need to provide milk and meat safely to their populations in the leased territories. Thus their efforts focused on protecting their dairy cattle and on securing meat through the construction of a modern slaughterhouse that invoked specific temporal and spatial controls on human and nonhuman animal behavior. Within the slaughterhouse compound, detailed times and designated spaces were established, including where and when livestock could enter and be taken to slaughter halls and how long the process should take. Temporal and spatial controls also extended to post-slaughter behavior, the use of machinery, and the deployment of personnel.

More broadly minded, the Japanese were primarily interested in the benefits that beef exports could provide in strengthening Japanese bodies back home. Inheriting the German management system, the Japanese tightened controls but extended them to spaces beyond the slaughterhouse compound to include controls on the borders between the compound and the stockyard. Moreover, since the export of beef initially involved the export of live cattle, the Japanese expanded the management system in the colony, intervening more directly in bovine bodies through sera and embedding them in strict temporal and spatial structures that highly protocolized the inspection process. Such monitoring activities as the Germans and the Japanese carried out were costly: the initial investment in building the slaughterhouse, the personnel needed to run it, the maintenance of the machinery, the complicated process of multiple inspections, and the long quarantines for live exports. All of these expenses were possible only with the financial support of the state, which underwrote the management activities and appointed a special group of scientific experts to act as agents. Foremost among these were veterinarians, whose knowledge of animals and their bodies, both living and dead, became central to the construction of a new kind of human-animal relationship in East Asia. Traders, farmers, or butchers could no longer choose how to interact with animals. Even slaughterers' interaction was embedded in scientific and technological management systems that disciplined and constrained their actions through adherence to detailed protocols. Many of these protocols have remained in place but are largely hidden from public view, even as other types of technological systems have transformed some animals into objects of visual consumption.

CHAPTER 3

TECHNOLOGICAL HUMAN-ANIMAL ADAPTATION AMONG THE PASTORALISTS OF INNER MONGOLIA

AURORE DUMONT

In the collective imagination, nomadic societies have supposedly never been penetrated by technology, the latter being habitually equated with Western and modern industrial, digital, and high-tech innovations. Nonetheless, throughout the twentieth century major inventions like cars, industrial infrastructure, new breeding techniques, and electronic devices have transformed the production system and livelihoods of millions of people, including pastoralists. Even though "rampant technological, economic, and political destabilization have nearly wiped away a mythical past that was once central to the self-image and world view of anthropologists," the popular representation of timeless nomads moving across vast grasslands with nothing more than their herds and mobile dwellings dies hard.[1] Contrary to prevailing stereotypes that portray the nomadic lifestyle as "immutable and unchanged for centuries," pastoralists today use countless "modern" technological devices in their everyday lives and herd management.[2] These include, among other things, motorized vehicles, mobile phones, GPS, and other digital technologies. Modern technology is thus an integral component of contemporary nomadic pastoralism in most places around the world and has greatly influenced the way pastoralists interact with their animals and environment. Nevertheless, nomadic pastoralists have not abandoned their own technology, which they combine with so-called modern technology.

This is the case in the Inner Mongolia Autonomous Region (People's Republic of China), where the minority groups who still practice nomadic pastoralism possess various skills, techniques, and beliefs that allow them to sustainably use the environment and to maintain a respectful reciprocity with their animals. Finding fresh pastures for herds requires advanced knowledge of the surrounding natural environment, from checking the nutritional quality of grass to finding proper water sources. It also implies taking into consideration the accessibility of the area, mainly in accordance with sociopolitical and spiritual factors. For those who combine herding with hunting activities, chasing a sable in the taiga is not only a matter of handling a rifle but also involves mastering numerous tracking techniques, such as identifying sable footprints and catching the game with the help of hunting dogs[3] and, sometimes, sacred entities. This expansive set of techniques, abilities, faith, and appropriate technologies together form what many scientists have called traditional ecological knowledge (TEK) or Indigenous science (IS). While some authors consider TEK to be a subsection of IS,[4] others use the terms interchangeably, as this chapter does. We will follow Inglis's definition of TEK as "an intimate and detailed knowledge of plants, animals and natural phenomena, the development and use of appropriate technologies for hunting, fishing, trapping, agriculture and forestry, and a holistic knowledge or 'world view.'"[5] TEK is not only "living knowledge"[6] made up of personal and collective experiences encompassing autochthonous/Indigenous[7] technological innovations suitable for the herds and for the economic, natural, and supranatural environment; it also has the ability to incorporate allochthonous technology (often thought of as officially sanctioned technology), thus creating interconnectedness and reciprocity. Indeed, as Berkes has noted, each knowledge system is legitimate in its own right, and the two kinds of knowledge may be pursued separately but in parallel, enriching one another as needed.[8]

This chapter is dedicated to the evolution of human-animal interactions through the prism of technology among Tungus[9] and Mongol pastoralists of Inner Mongolia. More precisely, it examines how the use of different technological items has impacted pastoralists' skills and mobility in relation to their herds from the beginning of the twentieth century to the present. Why and how have pastoralists adopted some technological devices and techniques in their herd management while rejecting others? How can allochthonous technologies integrated locally by national policies be given a place in the nomadic pastoral system? The exploration of TEK allows us to better understand how pastoralists have been innovating with both forms of technologies and how they, together with their animals, have adapted to the changing economic and sociopolitical environment in China. I argue that pastoralists have adapted their skills and practices to an ongoing process of technological change and that autochthonous technologies and other forms of so-called modern technology have been interacting with each other in a complementary manner for decades.

This research is based on ethnographic fieldwork conducted between 2008 and 2019 in Hulun Buir, a vast pastoral area situated in the northeastern corner of the Inner Mongolia Autonomous Region between Russia and Mongolia. In order to understand how allochthonous and autochthonous technological systems work together and affect the ways in which people live and work with their domesticated animals, I practiced participant observation among the Tungus and Mongol pastoralists, conducted interviews, and had informal discussions with them during their daily routines.

Hulun Buir's rich resources attracted Tungus and Mongol nomads, who settled in the area between the mid-eighteenth century and the 1930s. Today Hulun Buir is Inner Mongolia's most multiethnic area and has always had a distinct identity from the rest of the region.[10] Regarding its ecology, economy, and demography, Hulun Buir can be broadly divided into two ecosystems. With its typical pine, birch, and larch forests, the Siberian taiga runs from north to east; here small groups of Tungus people are traditionally engaged in reindeer herding, irregular hunting, horse herding, and farming. In the southwest the forest gives way to grass-covered steppe, where most Mongols and Mongolian-speaking Tungus[11] live. The steppe area is favorable for herding multispecies livestock, commonly known as the "five muzzles" (namely camel, horses, cattle, sheep, and goats). People differentiate them as "long legs" (camel, horses, and cattle) and "short legs" (sheep and goats) according to their appearance and behavior. The analysis of both pastoralist systems allows us to compare how technology is used by local people and how it affects pastoral management in two different ecological environments.

This chapter is organized chronologically. The first section provides an overview of the evolution of selected Indigenous technologies at the beginning of the twentieth century. It shows that every item produced by pastoral techniques and local knowledge are themselves a form of technology adapted to sustain human-animal symbiotic relationships. The second section analyzes the large-scale allochthonous technologies introduced by the Chinese state into the nomadic economy in the mid-twentieth century. It shows how government policies also used Indigenous technologies to serve national production and how relationships between men and their herds were reshaped through this new technological paradigm.

The last section explores the way motorized vehicles and dwellings have gradually replaced domesticated pack animals in the 2000s and altered pastoral mobility. It highlights a parallel process of limitation both in time and space. I show that, as human mobility has been reduced within nomadic space, human-animal sociality is also in decline.

INDIGENOUS TECHNOLOGIES AT THE BEGINNING OF THE TWENTIETH CENTURY

While crossing Hulun Buir in the first half of the twentieth century, many travelers, ethnographers, and officials from different countries described the Tungus and Mongol nomadic lifestyle in travel and political narratives.[12] These observers were particularly interested in depicting items used by Indigenous people in their "traditional activities," namely livestock herding, wild game hunting, and nomadization, a set of activities all linked to animals. The most accurate accounts not only presented the informal aspects of nomadic artifacts, but also offered detailed descriptions of the numerous techniques used in these peoples' herd management and the way they acted on their animals and environment.

Among the documentation available on nomadic artifacts and practices, I have intentionally selected those where ethnographers presented their subjects as "traditional," that is, made by local people with their skills and materials from the natural environment. This choice serves two purposes. First, it shows that Tungus and Mongols possessed their own technologies, and second, it makes it possible to analyze how these Indigenous technologies disappeared, were transformed, or were adjusted in accordance with the penetration of allochthonous technologies throughout the twentieth century.

During his Manchurian expedition from 1915 to 1917, the Russian anthropologist Sergei Shirokogoroff encountered a group of a few hundred Evenki people in the taiga.[13] They lived off a combination of game hunting and reindeer herding. Shirokogoroff explored what he called Evenki "technical adaptation," the various techniques the people acquired from their primary milieu and animals, as well as the manner in which they adjusted them to new techniques brought by allochtons.[14] According to Shirokogoroff, one of these was the "Tungus system of communications" (or "paths"), which he described: "To the eyes of people accustomed to railways and artificially erected high-roads with bridges and dams, the system of Tungus paths would not seem to be a technical achievement [or] a cultural adaptation. However, this is not so when one looks more closely at the phenomenon."[15] This "system of paths" refers to a complex technological system encompassing environmental sustainability, mobility skills, and animal domestication techniques. Following the annual renewal of resources, the Evenki moved along regular nomadic routes, selecting a new campsite according to the variables of each season. In the winter, the high season for hunting, the nomads pitched their tents wherever game was abundant, whereas in the summer they sought soils rich in lichen and water sources for their reindeer herd. The whole social unit traveled together with its herds every two to three days during the winter and every ten to twenty days during the summer. Once the most favorable area had been chosen, the Evenki set up their new camp in an area marked out in advance by the men, who opened new paths by cutting away branches. Extensive use of territory and a high level of mobility were thus crucial for both the reindeer, whose seasonal food required regular renewal, and the hunter, who, following the migration and reproductive cycles of species, frequently changed hunting grounds in order to increase his chances at catching game.

In order to support mobility for herd management, the Evenki devised a series of inventions, such as saddlebags for reindeer, canoes, skis, and movable dwellings. As both physical objects and social products, they were the results of technological action (finding and choosing material, cutting, softening, assembling, etc.) combined with technical and social functions. This supports Sneath's definition of pastoralism as "a sociotechnical system of activity that links techniques and material objects to the social coordination of labour."[16]

These autochthonous technologies also supported familiarity and reciprocity between men and reindeer or, more precisely, what Beach and Stammler have called "symbiotic domestication." The latter is one concrete expression of an animistic worldview wherein human and animal "persons" are conceived as equals in a reciprocal, symbiotic relationship, not only for their movements across the landscape, but also for their very sustenance and reproduction, their lives and deaths.[17] Domesticated reindeer indeed played a crucial role in Evenki mobility and their "system of paths." Each nomadic family usually kept small herds of domesticated reindeer[18] (oron in Evenki) and used them for milking, riding, and carrying loads. The reindeer were slaughtered only for weddings and shamanic rituals or in cases of incurable disease. Well adapted to the forest and able to move across marshy and snowy ground, the reindeer were one of the most indispensable means of transportation in everyday life: "With a good reindeer, one may travel over 50 miles a day while the horse cannot do more than 35 miles."[19] Evenki rode reindeer on various journeys, such as hunting trips that could last two or three days, visits to the villages for trade, and nomadization from one camp to another. Able to carry up to 50 kg, adult male reindeer were loaded with various harnesses, saddles, and saddlebags adapted to both the animals and the herders.[20] Using reindeer as transportation was the result of the daily interactions herders had with their animals. Packing, loading, leading, and riding a reindeer were all different technologies of domestication that allowed people to be mobile and "to follow the reindeer that follow the desires of humans."[21]

Birch bark canoes and skis were the other items used for transportation within the forest. "The canoe was mainly exploited during the summer season for hunting, crossing rivers and when they [Evenki] went to visit the banks of the Argun and Amur rivers."[22] The hunters preferred birch bark canoes for hunting the red deer often found feeding near rivers. "The hunter may get very close to the animal without any noise in a small canoe for one person."[23] Pine wood skis called kigle were also used for hunting deer and moose from November to February or March. When the weather was good, one could ski up to 80 km a day.[24] Requiring precise production techniques, items used to facilitate human movement were mainly manufactured with materials available in the surrounding environment. For instance, birch bark, carefully selected from among many local varieties, could be used only when soft, which involved a long process of steaming and various other technical operations.[25]

FIGURE 3.1. The *dju*, the mobile dwelling used by the reindeer herders in the taiga. Photograph by Aurore Dumont, December 2008.

Furthermore, each Evenki nuclear family possessed a mobile dwelling designed to facilitate frequent movement, easier packing, and transportation on the back of the reindeer. Called *dju*, these were conical tents made with wooden poles and covered with birch bark during the summer and skin or cotton during the winter. The *dju* was light and easily transportable by reindeer during frequent changes of location. The mobile dwelling was not just a habitat: it was also the center of herder economic, social, and ritual life.

In the southwestern part of Hulun Buir, the Mongol and the Tungus used the steppe as pastures for their five muzzles. Although all five provided milk and meat, each muzzle also offered specific resources. Sheep and goats were preferred for their meat and wool. As well as being beasts of burden, cattle gave meat and milk, which could be transformed into butter, cream, yoghurt, and cheese. Horses were valued as riding animals and were initially used only in the Mongolian "three manly games" (usually called *naadam* and made up of horse racing, archery, and wrestling). As the strongest beast of burden, the camel's employment as a pack animal made sense to pastoralists who had to move continuously between pastures. Nomadization routes were thus organized in all four seasons over a given territory, following the resources suitable for each type of domesticated animal. This extensive mobility was made possible thanks to several forms of technological transportation designed by the pastoralists. Among the different types of carriages, one of the most popular was a type of two-wheeled wooden vehicle. Drawn by a cow, it served nomadization, carrying wood, water, and milk for seasonal work.[26]

At the beginning of the twentieth century, herders kept the five muzzles all together on similar pastures, creating what Fijn defines as a "co-domestic relationship" in which the social adaptation of animals with human beings occurs by means of mutual cross-species interaction and social engagement.[27] In this framework technologies conceived by men are intended to support these symbiotic interactions.

The pastoralists of the steppe opted for a circular tent covered with felt or willows. Called a "yurt" by Westerners, its Indigenous appellation is *ger*, which means "home." The ger was one of the first and most visible forms of nomadic technology on the steppe. As noticed by Carruthers and Miller while traveling on the steppe, "a yurt, owing to its shape and construction, is of all tents the most perfectly adapted to withstand wind."[28] Setting up the ger necessitated choosing a suitable location and assembling the structure's various components. Although materials and shapes could differ between places, the ger's structure remained quite similar, mainly consisting of a cylindrical frame composed of wooden slats and a felt cover, with a single door facing south or southeast. The components of the ger were all based on know-how techniques. Up to the 1950s the ger was the only mobile residence possessed by the Tungus and the Mongols of Hulun Buir and represented the basic social unit in an encampment, a place for ordering social and gender relationships as well as daily work. The female sphere of responsibility was centered on household tasks inside the ger (washing, cooking, childcare), while the male sphere was linked with tasks associated with animal herding outside the ger. The mobile dwelling was also a component of the nomadic encampment where people and their five muzzles lived with each other, inhabiting a shared ecological and social landscape. Indeed, the ger, as a domestic technology, together with its surrounding environment provided a space in which every category of person or animal could be located. In this manner it became a microcosm of the co-domestic relationships between people and their animals.

The pastoralists also made frequent use of hay mowing in the nineteenth and twentieth centuries in order to feed livestock during the winter, especially severe ones (*zud* in Mongolian). Haymaking contributed to a new way of managing the pastures and redrew patterns of nomadic mobility and herd management. The steppe was henceforth divided between pastures devoted to haymaking and those dedicated to grazing. Haymaking rapidly became one of the basic tasks of seasonal labor. It was performed by men from the end of July or August to October. Haying techniques underwent successive transformations, mainly under the influence of new patterns of progressive breeding brought by Mongol Buryat refugees from Russia to Hulun Buir after the October Revolution.

The first and most common hand tool used for haymaking was the scythe, with handles 152 cm long and blades 67 cm long. During a productive day of labor with a scythe, 150 *pudi*[29] of hay could be mowed. In the 1870s, while they were still living in Russia, the Buryats adopted the so-called Russian scythe, made of thin steel obtained from Russian farmers.[30] However, one of the major technological innovations at the beginning of the twentieth century was the introduction of horse-drawn haymaking in Hulun Buir by the Buryats. The Buryats themselves started using this machine under Russian influence in the nineteenth century before introducing it to China, where other nomads adopted it.[31] As Khazanov has shown, nomads did not live in a state of isolation and had frequent contacts with neighboring sedentary populations, from whom they borrowed several techniques and devices that they adjusted to their own needs.[32] The hay was mowed with two horses; one day of haymaking necessitated four horses in total. The horse, one of the five muzzles and traditionally used as a rider's ally, became integrated into haymaking. While the best horses were kept as riding and racing animals, others became beasts of burden, thus requiring new techniques of domestication and leading to a change in the way people interacted with their domesticated animals. The success of horse-drawn haymaking was linked to its ability to improve productivity: in one day, 10,000 kg of grass could be mowed.[33] In certain areas, the mowing of hay even became a seasonal profession, with grass-selling constituting the most important source of income for some families.[34]

Whether living in the taiga or on the steppe, local Tungus and Mongol societies at the beginning of the twentieth century shared common practices in their everyday mobile lifestyle and relationships with domesticated animals. Both reindeer and five muzzles herding required extensive use of territory in order to generate optimal conditions for domesticated animals or catching the wild game that were the main form of subsistence. Nomadic economies were correlated to the use of various technologies that people had created, changed, and adapted over decades. Furthermore, Indigenous technology encompassed different items, materials, and skills that were essential for both pastoralists and their animals. In that sense technology not only served the people in their daily tasks, but also supported the reciprocal relationships consisting of familiarity, close cooperation, and partnership between humans and their animals. The foundation of the People's Republic of China in 1949 was followed by various policies and the introduction of allochthonous technology that gradually reframed these symbiotic human-animal relationships.

THE CHINESE STATE AND THE INTRODUCTION OF "COMMUNIST TECHNOLOGY" IN THE 1950S

Throughout the twentieth century nation states launched systematic, large-scale technological projects aimed at improving production and social conditions. Gradually penetrating into every sphere of social life, these macro-level technologies affected the nomads' way of life and their relationships with animals. Indigenous technologies were adapted, transformed, or disappeared at the same time, after coming into contact with new tools and machines.

In the 1950s the new Chinese communist regime aimed to integrate the population within the newly established planned state economy. In the first instance, inspired by the Soviet model applied in Russia and Mongolia, the Chinese economy included socioeconomic policies driven by the concept of "modernization" (xiandai hua).[35] As Cao notes, "One of the notable features of modernized society is the rapid advancement of science and technology, whose achievements are quickly and substantively transformed into direct productive forces, leading to a rapid increase in industrialization and automation."[36] The Chinese state sought to "modernize" the peripheral areas of the country and its Indigenous populations by developing large-scale technological infrastructure we shall call "communist technology."

In the forests and steppe of Hulun Buir, communist technology rapidly penetrated the Tungus and Mongol way of life. The integration of this ideological and economic model was based on the imperialist idea that the modern technology transferred by the communist regime would help local societies and their "traditional economies" escape backwardness and inequality. Indeed, nomads were perceived as having "a low technological capacity and being dependent on the outside world."[37] However, despite the alleged incompatibility between the "traditional" nomadic way of life and modern communist technology, Indigenous technologies were not entirely banned by the new regime. On the contrary, they were soon carefully used as a springboard to incorporate new technological models that would serve national production. Over the decades Indigenous technologies went through various adaptive processes—disappearance, addition, transformation—that generated innovative herding and hunting techniques, as well as new relationships between people and their herds.

From the 1950s the systematic construction of large-scale infrastructure throughout the country reconfigured economic and environmental features. In Hulun Buir the forests and steppe were gradually connected with roads, railway networks, modern transportation facilities, and factories. Meanwhile hunters and herders had to follow a new economic model, the aim of which was to raise productivity and increase livestock numbers.

In the taiga at the beginning of the 1960s, recognizing the economic potential of reindeer antlers as an ingredient in traditional Chinese medicine, the local government collectivized reindeer on the model of the Soviet sovkhoz. Each year antlers were cut and sold to the state in exchange for work points. This is what Vitebsky called "production nomadism," referring to the treatment of reindeer under the Soviet regime. The animal was isolated from its social base and reduced to an instrument of production.[38] Evenki elders say that before the 1960s antlers were not cut because it caused the reindeer pain and could bring disease. The Evenki had to adapt to the new rules by creating tools and developing cutting and drying techniques. They devised a knife especially designed for antler cutting that caused only a minimum of pain to the animal, and men trained themselves in the use of this new technique. The communist idea of technological progress was thus intimately linked to the notion that extensive production (more furs, pelts, and antlers) could be applied everywhere, with some changes made according to the local environment.

On the steppe the five muzzles were perceived by the regime as species that could be raised separately and multiplied rapidly with the integration of new technologies. Furthermore, while reindeer could provide only antlers to national production, the five muzzles had more to offer, such as meat, skin, and wool. From the 1960s the composition of livestock changed significantly. In Soviet Russia and Mongolia the governments introduced foreign livestock breeds to raise productivity.[39] In China the number of sheep and goats grew dramatically to support the intensive production of meat and wool,[40] and pastoralists were encouraged to give priority to these two types of muzzles, thus redefining their herding practices. Collectivization, combined with the spread of allochthonous technology, coincided with the introduction of the specialized herding of a single species at the expense of traditional multispecies herding. Some people became specialized in one or two species: herding sheep, goats, horses, and camels became predominantly a pastoral job, with one herder being the only person in charge of domestication and taming, thus breaking the link that used to tie the entire nuclear family to the five muzzles.

In addition, the construction of macro infrastructure and the transformation of the environment went hand in hand with the systematic implementation of various micro technologies, with the "smallest" forms of allochthonous technology suiting the larger ones. Between 1953 and 1956 the Chinese government invested 18.5 million yuan in pastoral regions, according to official sources, in order to construct roads, buildings, and wells.[41] The regime engaged in a program of what was known as technological modernization, including the use of fences, the introduction of new breeds, and the construction of heated winter sheds, fences, and mechanized machines. A triangular technological process combining animal crossbreeding, haymaking, and mechanization was sponsored by the state. Changing the nature of the animal by feeding it in a new way with different mechanized tools was conceived by the Chinese state as the core of a new technological process for increasing productivity.

Already practiced in the pre-communist period, haymaking became a key element of extensive herding from the 1950s on. Haying areas were expanded to supply fodder for the animals during winter. It also contributed to longer stays in newly constructed villages during the winter. The larger quantities of hay were made possible thanks to the introduction of mechanized haying machines and tractors in the 1950s, which definitively replaced scythes and horses. The policies of the Chinese government were driven by the idea that highly mechanized pastoralism was synonymous with progress, modernization, and livestock raising. Mao Zedong's famous maxim, "Farming's fundamental opportunity for advancement is mechanization," was applied in the same manner on the steppe. The modernization of rural zones had to be accomplished with the mechanization of herding, completing the other large-scale infrastructure projects already implemented. Among the Khamnigan, a Tungus people engaged in five-muzzles herding, the mechanized haying machine was introduced by the government in the 1950s and could cut the same amount of grass as twelve to fifteen people using scythes.[42] Horse-drawn haymaking was definitively abandoned in the 1970s.[43] The use of various mechanized machines pushed the pastoralists to learn new grass-cutting techniques and to develop the skills required to properly use tractors. They adapted their seasonal task to this new vehicle. The triangular technologies also gradually changed the way the herders interacted with their herds. As was the case with reindeer in the forests, steppe animals were an instrument of production whose social function was lost. Lifelong association with animals was gradually altered along with autochthonous herding techniques.

From the 1960s on villages and roads were considered key infrastructure in Hulun Buir; they played an important role in the way the state reinforced its visions of change and progress in peripheral areas. Both in the forest and on the steppe, villages for the Mongol and Tungus minority groups were constructed with schools, dispensaries, and fixed wooden and mud-brick houses. The sedentarization of nomadic societies into newly built villages thus reflected the modernist vision of the Chinese communist project that implied that settled life was more advanced than traditional nomadism.[44] The growing number of settlements and roads in pastoral areas impacted pastoralists' lifestyle, mobility, and relationships with animals. However, the construction of settlements and fixed houses was not incompatible with pastoralists' mobility. Indeed, as livestock required regular moving, sedentarization was not totally complete. The government encouraged what it called being "settled with moving herding": while some members of the family remained in the village (mostly the elderly, school-aged children, and women), others still moved with their herds across the grazing lands.[45] In the late 1980s, nearly thirty years after the sedentarization policy, Sneath noted relative stability in Hulun Buir's social organization, the division of labor, the symbolic sharing of encampment space, and nomadization. In this steppe area more than half of the Mongolian pastoralists did not have fixed dwellings and came to the village from time to time with their yurts.[46] In contrast, in some other areas of Inner Mongolia people have abandoned the circular yurts to construct rectangular fixed houses similar to farm houses.[47]

The construction of road infrastructure changed the features of the forest and the steppe, and thus pastoralists' mobility. The nomadic landscape became partially divided into lanes, paths, and routes in order to facilitate the transportation of people and goods. Local people had to follow new trajectories between their nomadic spaces and freshly created sedentary spaces. At the same

FIGURE 3.2. The modern rectangular tent *zhangpeng* used in the taiga. Photograph by Aurore Dumont, July 2016.

time, the development of industries led to the arrival of many Han Chinese migrants. Furthermore, while roads and adjacent infrastructure for transportation were growing, new types of mechanized and motorized vehicles, such as tractors and trucks, appeared to serve local production. However, the nomadization between nomadic camps was still mainly conducted by reindeer in the forest and by camel, horse, or horse-drawn carriage on small unpaved roads on the steppe.[48] Domesticated animals were thus not only a source of income and social value, but also an essential technological companion for transportation. This required a mutual process of adaptation between men and their herds.

Furthermore, the nomads of Inner Mongolia also had to adapt their nomadic dwelling technology to new types of dwelling, both fixed and mobile. In the late 1980s the Evenki conical tent (*dju*) living in the forest was gradually replaced by a rectangular tent known by the Chinese term *zhangpeng* (meaning "tent"). The shape of this rectangular tent is reminiscent of a fixed house found in a village. The adoption of the rectangular tent has led to adjustments in the various techniques previously used by the nomads to construct their nomadic dwellings. Everything is made in the workshops of Hailar, a city specialized in the manufacture of ethnic goods; the only part of the tent that is still constructed by the people themselves is the frame. Over time the Han Chinese have furnished these mobile homes with additional equipment. Larger and lower, the latest model tent has two windows on each side (instead of one) with canvas shutters; the chimney is now located at the back of the tent instead of in the middle. As the external architecture changed, the contents of the domestic interior have gotten heavier, making nomadization impossible without motorized transportation.

On the steppe the yurt has also undergone significant changes in terms of the materials needed for its manufacture and the layout of its space. Traditionally the yurt consisted of a cylindrical frame made of wooden slats and a felt covering, with a door often covered by painted wood. Formerly made by families through their knowledge of the relevant skills and techniques, these components are now made of metal and plasticized canvas, the latter now being preferred to felt since it is less expensive and easier to find. The opinion of pastoralists regarding the use of these new materials is ambivalent. All believe that metal is inappropriate in the summer because it attracts heat (unlike wood, which maintains freshness). Furthermore, the allochthonous materials used are not necessarily adapted to the animals. Indeed, the new dwellings do not allow people to hear the animals well enough in cases of distress. On the other hand, they highlight the convenience of this dwelling and say that they have more time for other activities now that they can purchase ready-to-go yurts.

The shift to settled housing has also altered gender relations. For a significant part of the year, the traditional responsibilities of women in the nomadic encampment have become male tasks, since women need to stay in the villages to be with children attending school. The use of new dwellings also brought great changes in the way pastoralists interact with animals. People now living in new sedentary spaces have gradually lost their herding skills and, most important, the symbiotic familiarity they used to maintain with their herds. On the other hand, social engagement between pastoralists and animals in nomadic spaces has been maintained and sometimes reinforced.

CHANGING PASTORALIST MOBILITY AND HUMAN-ANIMAL SOCIALITY IN THE 2010S

From the 1960s various mechanized items became indispensable tools in herd management in Tungus and Mongol societies, even though animals were still used as means of transportation. The tractors that replaced horse-drawn haymaking on the steppe are an example of the adaptation of autochthonous skills to allochthonous technology. While the implementation of the first mechanized items was carried out by the state on a large and collective scale, nowadays the use of motorized and individual vehicles is a matter of personal choice. In the late 2000s motorized vehicles became affordable for most nomadic families, and their systematic use has had a deep impact on the way people interact with their animals and move with or without them.

In the 1990s and especially in the 2000s, while motorcycles, cars, and trucks became common for pastoralist families, animals lost their pack and transportation function. In Mongolia Fraser has demonstrated that technological change may be understood through the concepts of *deskillment* and the *transposition* of skills. While "deskillment" describes the processual loss or forgetting of skills in tandem with processes of change, the "transposition" of skills refers to the embodied movement of skills from one context to another.[49] This approach suits our Inner Mongolian case, where the adoption of motorized technology led to a transposition of herding skills and mobility and a reshaping of human-animal sociality.

The spread of motorized vehicles among nomads is the result of multiple factors, one of which is the development policies carried out by the Chinese state from the beginning of the twenty-first century. In 2000 the state launched the Open Up the West (*Xibu da kaifa*) policy. This project aimed at "encouraging economic growth, to reduce socio-economic inequalities, and to ensure social and political stability in non-Han areas of the PRC."[50] Still shaped by the notion of modernization, the campaign sought to bring strong socioeconomic and technological development to peripheral areas inhabited by minority groups. In Hulun Buir new provincial and national roads were constructed, while small roads were paved deep into the forests and steppe. The creation of a systematic road network supported goods, tourists, mining exploitation on the steppe, and also herder mobility. In parallel, pastoralists' living standards improved as they gained access to credit and started to buy motorized vehicles.

Motorcycles, cars, trucks, and trailers are the allochthonous technologies that have generated new ways of being mobile, as well as innovative interactions between people and their herds. Motorcycles soon became the ultimate motorized vehicle, thanks to the affordable price and the fact that they did not require a driver's license. Used by both men and women, they are employed to connect the sedentary space (the village) to the nomadic space (the camps or the pastures) in a quicker and easier way. Iselin has shown that among the Tibetan pastoralists of Sichuan Province, translocal motorized mobility practices facilitate the expansion of space and place, although this space is fragmented.[51] In Hulun Buir pastoralists ride motorcycles during the summer from their village to their camps or pasturelands and also within a single space (for example, the village) to visit relatives or for festive occasions.

Motorcycles have also become an adapted technological tool in herd management. Today, most herders ride motorcycles on the steppe to move animals (especially cows, sheep, and goats) from one pasture to another, to bring them back, and to look after them. According to people, this allochthonous technology provides an efficient and quicker way to control and follow animals. Animals react positively to the noise and movement of motorcycles because they are used to being led by men on horseback. In this manner, men, sheep, and goats have

FIGURE 3.3. Two reindeer herders on a motorcycle. Photograph by Aurore Dumont & Alexandra Lavrillier, July 2014.

maintained mutual engagement through their respective movements. When pastoralists transpose skills associated with horse riding to the motorcycle, they not only meet technological change through the adaptability of their autochthonous skills but also adjust to their animals, which in turn develop habitus fitting the new technology.

Trucks and tractors are the other vehicles that have become indispensable in nomadic space. Their systematic use began in the 2000s in order to facilitate herd management and to carry heavier loads over longer distances. Since then, the truck has become a multifunctional vehicle, mostly used for carrying livestock and furniture between nomadic camps or from a camp to the village. The truck works as a substitute for pack animals, as it has totally replaced reindeer in the taiga and horses and camels on the steppe during nomadization. Before, animal-drawn transportation required the embodiment of autochthonous skills, such as the use of ox or camel carts, and deep knowledge of migration routes. Since motorized vehicles were introduced, both men and animals have gone through a process of *deskillment*. The knowledge and skills previously used to pack an animal have been lost, causing mutual disengagement between people and animals during nomadic mobility. One reindeer herder explained to me that while he could ride and pack a reindeer easily in the 1960s, this is impossible for him today. Not only has the reindeer's body changed so that it is no longer adapted to carrying things, but the animal's behavior has also shifted; no longer trained as a beast of burden, it is unsuited to packing technology.

Through the acquisition of motorcycles, cars, and trucks, pastoralists are able to make more regular excursions to their camps while also developing new driving skills adapted to their forms of movement. Motorized vehicles are not only convenient for transporting people, goods, and animals; they also offer new ways of being mobile within a definite space. While movement determined by herding (movement from one camp to another within nomadic space) has been reduced in recent decades due to socioeconomic transformation (sedentarization, the fragmentation of nomadic routes, and territorial restrictions), pastoralists' mobility within various types of space has paradoxically increased, since motorized vehicles carry goods and people over longer distances and at greater speeds. As pointed out by Humphrey and Sneath, "Mechanization can be used to support the distinctive Inner Asian tradition of high pastoral productivity through mobility."[52]

The nomads of Hulun Buir live between sedentary and nomadic spaces, alternating between mobile dwellings and fixed houses and adjusting their social practices, herd management, and technological skills to these different dwellings. For example, among the Barga Mongols of Hulun Buir, Marois has shown that, by moving from a yurt to a fixed house, people modify their relationship with the environment, encouraging new and exclusive forms of intimacy,[53] both between people themselves and between people and their herds.

FIGURE 3.4. A trailer on the steppe. Photograph by Aurore Dumont, July 2019.

In the 2010s a new type of mobile dwelling appeared on the steppe and in the forest, once more reshaping pastoralists' mobility. Called a "trailer" (*peng-che* in Chinese and *muilag tereg* in Mongolian), it serves both as a vehicle and a dwelling. It is halfway between a rectangular house and a vehicle mounted on wheels.

Like the yurt, its door is oriented south or southwest. The trailer always comes with a small tractor, which is required to move it. Although it was adopted less than ten years ago, the trailer is used throughout the year, in contrast to tents and yurts, which are mainly used during the summer. The trailer has become an everyday vehicle in pastoralists' lives and is the most visible vehicle on the steppe. It was introduced by the Chinese government and promoted as a "modern technological tool" for herd management and mobility. The original concept of using a movable wagon to travel across the countryside was first experimented with among Chinese farmers under the government concept of a "mobile trailer for long-distance education" (*yuancheng jiaoyu da pengche*), whose aim was to give schoolchildren and students access to long-distance learning even in the most remote areas of the country. Adopted in farming areas, the long-distance education program was implemented by local authorities to provide the rural population with additional information on farming techniques, new policies, and other information concerning the life of the village. In 2011 the Chinese government extended this concept to the steppe areas of Hulun Buir (and in 2013 to the forest areas) "as part of a program aiming at reducing poverty and supporting pastoralists' mobility."[54] As is often the case, a technological and governmental device that works in one locality is quickly adopted in the rest of the country. For example, among the Mongol Barga the Discipline Inspection Board set up a government trailer that went across the steppe to meet with the pastoralists. The primary goal was to ensure control over the territory and its population. In the words of the local authorities, the trailer serves as a "bridge between the local government and the often dispersed pastoralists."[55] The second objective is to inform the local people through educational activities of new government policies and to transmit ideological messages. By 2018 the trailer, as the motorcycle and the tractor, became an indispensable technological tool in pastoralists' lives; every household possesses one.

While trailers possess multiple functions, the main one is herd management, especially in sheep herding. Requiring extensive mobility, sheep herding relies on regular rotation between pastures. The five muzzles were traditionally herded together; however, over the last two or three decades sheep herding has become a specialty. For sheep herders, who habitually keep their herd together with other people's, the trailer has become an indispensable tool on the steppe. According to them, it is easy to handle, very convenient, fast, and allows sustained movement without any outside help. Indeed, while the yurt requires disassembly and reassembly, preferably by a team of several people, the trailer does not need any human intervention, since a simple tractor is enough to move it. With the technological specialization of sheep herding, pastoralists are now often single men or a couple and thus need a more convenient way to move within a territory. Its high price (between 20,000 and 30,000 yuan, or US$2,880 to $4,320), which is ten times more expensive than a yurt, does not seem to discourage pastoralists from acquiring one. Apart from sheep herding, trailers are used for haymaking in the autumn and have replaced yurts everywhere. In addition, they are used as a mobile vehicle and a dwelling, mainly by men. These trucks are also used within the village and adjacent spaces, where they serve as a warehouse or, more rarely, as a home. Today, pastoralists not only live alternately between fixed houses and mobile dwellings, but they have also adopted various kinds of mobile residences, using them according to their needs, pastoral work, and the changing seasons.

I have explored a century of technological human-animal adaptation among the nomadic pastoralists of Hulun Buir in Inner Mongolia. In both the forest and the steppe, local people share their own technological mastery based on environmental knowledge, manual skills, and day-to-day interactions with domesticated and wild animals. As we have seen, technology is an ongoing process in which some objects, behaviors, skills, and techniques disappear while others emerge or change, thus reframing the relationships pastoralists maintain with their animals. Technology may also be an interesting way to highlight resilience and adaptation in a given society and its animals and to capture the way technological factors interact in complex ways with socioeconomic spheres and government policies.

At the beginning of the twentieth century Tungus and Mongol nomadic societies conceived of their own technology in accordance with their herding and movement needs, maintaining at the same time a symbiotic relationship with their domesticated animals. The introduction of "communist technology" from the 1950s led to the appearance, transformation, and disappearance of specific autochthonous technological items, while the pastoralists had to develop various strategies and skills to fulfill new patterns of living, herd management, human-animal relationships, and mobility. In the 2000s the introduction of motorized vehicles affected patterns of movement, while pack animals and their related technological skills have disappeared. Finally, the constant evolution of mobile dwellings also highlights the ways pastoralists have adapted their skills and practices to an ongoing process of technological change.

Among nomadic societies, the issue of technology recalls the still ongoing debates about urbanization, sedentarization, and industrialization. What we call "modern technology" allegedly makes nomadic peoples inauthentic, since "traditional" knowledge is often opposed to modern technology. As Humphrey and Sneath have pointed out, the term "nomadism" is a category imagined by outsiders that brings with it various stereotypes about pastoral peoples, such as their peripatetic movements or "low" technological capacity.[56] However, as I have tried to show, there is no barrier between autochthonous technologies and other forms of "modern technology"; they interact with each other in a very complementary manner.

CHAPTER 4

NEGOTIATING HUNTING ETHICS THROUGH TECHNOLOGY

KARIN DIRKE

Taking life is never trivial nor a matter to take lightly. This fact seems to permeate the history of killing.[1] Most people need thorough training to be able to kill. Methods, as well as technology, have been developed continuously and expansively throughout history, however never as efficiently as during the nineteenth and twentieth centuries. During this time killing (of animals as well as humans) took on industrial proportions.[2] This chapter is concerned with the killing of animals in late nineteenth- and early twentieth-century Sweden.

Hunting has historically often been used as ritual war training.[3] Royalty practiced the close, face-to-face killing of individuals by having animals collected behind screens or nets and killing them at close range. During the nineteenth century hunting practices changed in Sweden and the rest of Europe. Hunting became a more structured and scientific part of forestry and the management of wildlife, as well as an increasingly popular pastime, conducted for amusement.[4] *Game management* became an important concept. This primarily meant sparing female animals to promote regeneration, feeding game animals during the winter, and killing predators. Turn-of-the-century hunters were more often individuals walking the woods, accompanied by their dogs and carrying their guns, looking to shoot animals for pleasure, as well as to learn about them. Toward the end of the nineteenth century hunting became a topic within the debate about animal welfare in Sweden. A new sensitivity to the suffering of animals had spread across Europe during the nineteenth century, and in Sweden the welfare of animals had been a topic of debate since the 1870s. Animal welfare societies emerged in Sweden at the time and were at first primarily focused on the welfare of animals close to humans, such as horses, farm animals, and pets. The animal welfare societies thus were mainly concerned with topics other than hunting, such as slaughter, transportation of animals, farmers' treatment of livestock, and, not least, animal experimentation. Toward the turn of the century, however, the death of animals, through slaughter or hunting, became a contested issue. Hunters now became increasingly aware of the criticism raised by the animal welfare movement against their unethical treatment of animals. Some hunters themselves began to question the large-scale killing of animals, putting entire species in danger.[5] Ecological issues had been raised in Sweden at the time, and by the end of the period studied here both a new animal welfare law had been passed in Sweden (1907) and the nature conservation movement had become organized.[6]

This chapter aims to investigate the language surrounding the technology developed and used in a late nineteenth-century context through the stories about hunting in the twice-yearly magazine *Jägaren* (1895–1907). Through a close reading of the hunting stories and an analysis of them relative to the endeavors of the contemporary animal welfare movement, the intention is to illustrate how the killing of animals was managed in relation to the growing anti-cruelty regime of the late nineteenth and early twentieth centuries.[7] *Jägaren* was quite typical of the kind of popular hunting stories published in Europe during the nineteenth century and was inspired by similar publications in Germany, England, and France. The journal will in this chapter function as an example of how ethics, ecology, and animal agency were negotiated in late nineteenth- and early twentieth-century prose on hunting.[8]

At the same time as *Jägaren* emerged, Swedish society was permeated by a rising new sentimentalism, which proliferated especially in the upper classes, for whom hunting was a popular pastime. Just as their counterparts in other parts of Europe, animal welfare societies in Sweden had close connections to the royals and nobility. The suffering of animals became an important topic of conversation in society during the late nineteenth century. By the beginning of the twentieth, however, hunting had entered a critical discourse, and hunters now found themselves having to defend their activities.[9]

The people active in animal welfare societies and the hunters involved with *Jägaren* were very much in the same social stratum, mainly the upper class.[10] They were concerned with the same topic, the killing of animals, and were striving to establish their accomplishments as ethical. They were concerned with the ethical treatment of animals and specifically how this should be achieved. Hunters argued, with society and each other, about the most ethical, humane way of killing animals. Often they referred to and used technological solutions to realize this.

The analysis of *Jägaren* will also concern the representation of agency and how it was ascribed to individuals, groups, or objects.[11] I believe that agency is often represented and distributed by humans when dealing with problematic aspects of society. This chapter discusses how this is done in a specific context: the late nineteenth- and early twentieth-century human-animal relations in Sweden, more specifically the killing of animals.

THE MAGAZINE *JÄGAREN*

Jägaren was published by Hugo Samzelius, a forest ranger, writer, and passionate hunter from Nyköping. Samzelius worked for long periods of time in the northern parts of Sweden as a forester and traveled as a collector and ethnographer for the Nordic Museum and for the open-air museum Skansen. *Jägaren* was a Nordic journal aimed at hunters in Sweden, Norway, Finland, and Denmark, and was published alternately yearly or semi-yearly, and was financed by being sold in bookshops as well as by subscriptions and advertisements. It consisted primarily of an illustrated collection of stories and poems about hunting and wildlife as well as biographies of reputable hunters. *Jägaren* was primarily written in Swedish with some articles in Norwegian and Danish. It was given up by the editor in 1907 because of his feelings of remorse over hunting. In the beginning of the 1900s Samzelius had increasing contacts with the animal welfare movement, which brought about a skepticism concerning the ethics of hunting practices. In the very last issue of *Jägaren* Samzelius wrote a farewell to his reader, stating that over the years he has become increasingly uncomfortable with hunting, especially hunting done solely for the purpose of enjoyment.[12] He writes that he has come to the conclusion that the Nordic landscape is so "admirable" and the sporting life so wonderful, that it seems "completely unnecessary to 'improve the pleasure of being in nature' by, merely for amusement, wandering around, shooting some perfect and beautiful animals."[13] Samzelius concludes that *Jägaren* has transformed from being hunting friendly to a position of pro-animal protection, a change that has made the editor's position difficult in several aspects.[14] It became increasingly difficult for him, as editor of the magazine, to publish entertaining stories about hunting while personally having doubts about the ethics of hunting. This interesting transformation of the editor's viewpoint makes it an especially suitable material for the study of the attitudes to killing animals. *Jägaren* contains several layers of passion: for nature and landscape, for traveling, and for animals.

THE STORIES

The hunting stories in *Jägaren* have similar traits and a comparable structure. Most often the narrator tells a story about how he travels through the landscape, relating who is with him and how the environment appears. Details about the surroundings, the weather, and the hardships are ubiquitous. The buildup of the story is often quite long, slowly escalating to the peripeteia of the story, which is centered around the hunter firing his weapon.

Jägaren was presented as a literary attempt to give Nordic sportsmen domestic hunting stories.[15] These stories are framed as travel accounts even though they often are enacted in the local area (around Stockholm). Some, especially the ones about hunting large animals such as bear or moose, take place in the north. The narrator, most often of the upper class, is a person who travels for scientific or tourist purposes. The differences between the narrator (his superior skill, his outstanding use of technology, his attractiveness) and the local population are often accentuated in the stories.[16] Nearly all anecdotes are told as remembered accounts, thus partly claiming some degree of truth, though the genre itself is characterized by a certain semi-fictiveness. Hunting is described as a pleasure but also for the purpose of collecting specimens for scientific reasons.[17]

Ethical guidelines are often embedded in the stories. It is obvious that, though the killing of animals is endorsed, it should be done in a special way. Hunting most often means shooting animals, not catching them with poison or traps. For example, in a story about traveling in Finnmarken, the hunter Gustaf Kolthoff is critical of the unethical trapping of animals. In the same volume of the journal, another hunter argues against the use of poison or traps, which he describes as methods of torture. Other stories are also critical about hunting methods or shooting too many animals, such as one criticizing a method of killing ducks that reminds the author of the "Italian populace and their exter-mination of migrating birds."[18] It is frequently mentioned in the stories that the author's hunting is performed within the regulations, abiding by the rules of when and where hunting certain animals is permitted in Sweden.[19] One of the specific aims of *Jägaren* was to spread information about hunting regulations and ethics. The suffering of hunted animals is occasionally addressed. In one of the few accounts of women hunting, a memoir by Frida Segerdahl-Nordström, the story ends with how the hunter shoots a moose twice to be sure to quickly put it out of its misery.[20] In the following biography of Segerdahl-Nordström, Samzelius notes that she is both a hunter and promotor of animal welfare. The two concepts are perfectly compatible, the editor states, "since in *real* hunting assassinations or cruelty to animals is not acceptable."[21]

In *Jägaren*'s hunting stories we are commonly provided with details con-cerning the weapons and ammunition used.[22] The model of the guns and caliber of ammunition used are regularly mentioned.[23] Information about the arma-ment of the hunter is of course used to either emphasize their skill and courage or to insert ethical decrees about the importance of using the correct weapon for the occasion.

Information about whether the narrator is carrying a gun or not while trav-eling through the landscape is almost always conveyed.[24] The information about the caliber of ammunition used also contains ethical meanings. When a hunter confesses that he "now as well as the last time, used shots, which were rather too fine for winter fox at a long distance, namely eng n:o 2," he is not only conveying information about what kind of ammunition he used, but also implying the risk of wounding the animal rather than instantly killing it.[25]

The technology of hunting weapons was greatly improved in Sweden and Europe during the nineteenth century. Front-loading guns were subsequently replaced by breechloaders, which were quicker to reload and easier to aim. Military weapons were developed and used in hunting until the beginning of the twentieth century, when military guns became designed for mass destruction and thus were forbidden in hunting.[26] In the hunting stories in *Jägaren* the efficiency of the weapon is often referenced, if not explicitly discussed. The guns used are most often the modern breechloaders, and the narrator compares his own modern rifle with the antiquated guns of the local population.[27] These older guns are described as imprecise, often wounding the animal rather than killing it. The stories thus become permeated by a modernization narrative in which the desired hunting practices are framed as ethical as well as modern. The ability to shoot flying birds is related to having a fast gun. Therefore the hunters in the stories refer to the skill of being able to work out where to aim at birds in flight, which requires knowledge of the speed of both the bird and the bullet, as well as their relation to each other.[28] Modern technology thus signified ethical hunting practices.

TIME IN THE HUNTING STORIES

The hunting stories of *Jägaren* are most often narrated in the form of memoirs. The narrator is recalling past events, as all stories are written in the past tense.

The stories display different layers of time and contain different, and sometimes conflicting, temporalities. The base of the story is the chronology of the hunt, the quite simple structure that frames all hunting stories. This style of narrative begins with the hunter traveling through the woods, escalates when the animal is encountered, culminates with the killing of the animal, and ends with the hunter/narrator returning home. Time is sometimes an essential part of the story. The slowness of the hunt for the animal is stressed. The time it takes to hunt it down is emphasized. The actual kill, however, is often expressed in another time frame. Details of the death of the animal (such as the anatomy of where the animal has been hit), which could not have been known to the narrator at the time, are frequently embedded in the stories. These anatomical digressions insert into the story another time frame, the time after the hunt, when the animal is butchered and its wounds are examined. This way both the ethics of the hunt and its technology are entrenched in the story.

The narrated hunting events are also framed in a fictional, folkloristic time, where animals are able to rejuvenate and the forest is an eternal resource that can be continuously harvested. The hunting events described in *Jägaren* are sometimes fictionalized to the end that one animal is killed and the next one hunted is described as being the same individual. In one story, for example, the narrator describes how he shot and killed a hare, after which he returned to the hunt, "but this time the Hare was unwilling to comply."[29] The tale about this successful, and failed, hunting expedition gives the impression that hunting is a sport where both competitors are playing a game that starts over and over again. The playful aspect of the hunt is emphasized, ignoring the perspective of the animal. Instead, the animal is instilled with agency. The game involves two contestants, both willing participants. The responsibility for the unfolding events is deflected from the hunter. Instead the stories are played out as a fictionalized game between the gun and the animal, with the narrator as a witness. The motif of hunting as a game is ubiquitous in hunting stories, underlining the idea of *fair play*, central to hunting ethics to this day.

THE SOUNDSCAPE OF THE HUNT

The hunting stories in *Jägaren* are intended to give the reader a literary experience of hunting. The sense of being in the forest, the sights, sounds, smells, and feelings experienced by the hunter are depicted in the tales. The sense most commonly alluded to is hearing. The soundscape of the hunt related in the stories gives the reader access to affective aspects. Sound is connected to emotions. As historian Stephanie Rutherford has pointed out, sound is important for the historian to address. She draws attention to how sound can be interpreted differently in various contexts. The wolf howl, in her case, is for the wolves a form of communication with a wide range of meanings. For the settlers in Canada, however, the howls had an entirely different connotation, permeated with fear and apprehension. She argues that sound is often not awarded enough attention in historical sources; the sounds of the environment are a significant part of history and should be investigated. However, sound disappears in the historical sources, leaving the historian to interpret silence.[30]

This is not the case in the hunting stories in *Jägaren*; the soundscape of the hunt is an essential aspect of the narratives. The sounds related are, however, clearly intended to give only one side of the story. The sounds of gunshots emphasize the fact that the hunter walks through the environment with a certain purpose. The gunshots establish the fact that the story is not merely a portrayal of the beauty of nature, of a picturesque landscape. The story may appear to be a travel account but is primarily something else. Sounds in the stories are distinctly directed toward one interest only. The meanings of and the emotions attached to the sounds of the hunt for the animal are never addressed. Though the animals hunted are often awarded agency in the stories, their feelings in relation to sound are completely absent. The sound of a gunshot is obviously linked to a dead animal. Nonetheless, in the stories the specific death of the animal is often skipped. The kills, as well as their gory details, are often swiftly passed over, as when Gustaf Kolthoff, zoologist and founder of the Biologiska Muséet (Biological Museum), and his travel companions encounter birds of prey in the north of Norway: "Suddenly I heard a gunshot and saw one of the buzzards fall from a considerable height: later I found him dead in the hollow below me."[31] Or, when Kolthoff sees the smoke from his friend's gun, he notices how the body of the cormorant hits the mountainside and then "a while thereafter heard the gunshot."[32] When another hunter, Erik E:sson Rålamb, recounts an especially dramatic, but failed, hare hunt, the story is related in terms of sound: "Bang—the first barrel blared. The Hare flew up high in the air, but came down on all fours. Bang—the other barrel—and away ran the hare with long leaps disappearing like a white dot in the dark." Instead of crying out "all's dead," the narrator was forced to call the dogs to put them on the trail.[33] The account is, of course, a story of a miss, a wounding of the animal. That aspect is, however, only implicit in the story.

The barking dogs are also an important part of the soundscape of hunting stories. The dogs signify the naturalness and joy of the hunt. The sound of their chasing the hunted animal is described as sweet music, lovely to the hunter's ear.[34] The different hunting dogs could be played out as a choir when they picked up the scent of the trail. The vocalizing dogs convey messages, all meaningful to the participating hunters: "Then Snapp caught on. Krut accompanied with his slightly hoarse bark, while Pipa shrieked like a bell and Ladda kept the bass. 'It most certainly is fox,' said the nearest man."[35] The meaning as it might be understood by the hunted animal is entirely ignored.

AVOIDING FATALITY, DISPERSING AGENCY

One of the more prominent features of the hunting stories in *Jägaren* is the way they pass over the actual killing of the animal. The death of the animal is almost always euphemized. The animals are "sent to kingdom come,"[36] "landed in the game bag,"[37] "dealt with,"[38] et cetera. As we have seen, the animal is frequently awarded agency and thereby represented as sharing responsibility for the unfolding events. As one hunter sums up the expedition of the day: "After a couple of hours two hares had stained the snow with blood."[39] Another one describes how two old male capercaillie "flew too close to the muzzle of the gun and ended up in the game bag."[40] In these examples, the responsibility for the death of the animals is taken away from the hunters. The hunter becomes detached from the kill and reduced to someone holding the gun or carrying the game bag. The stories often tell how the animal suddenly gets, acquires, or contracts shots.[41] Or, as one hunter describes the end of a long bear hunt: "Finally my double-barrelled gun settled the matter."[42] The agency is in this way transferred to the gun; the animal is not portrayed as being killed by the hunter, but by the gun. According to historian Joanna Bourke, ascribing agency to guns is a recurrent way of speaking about them. Military weapons are frequently given names and are described as agents acting independently.[43]

Most technology of death aims at distancing humans from the act of killing. In her study of violence and its deep-rooted social aspects, Bourke points out how words used in the military industry, euphemistic and underestimating, disguise the fact that weapons are all about maiming and killing. Though killing seems taboo in human societies, the language used to describe warfare and killing is often pleasurable.[44] In that same way the hunting stories are designed to be pleasing. The tales are not framed as controversial or critical. The story is told by a hunter, and the audience is expected to be hunting-friendly. Bourke points to different ways of speaking about weapons, all aimed at distancing or concealing the fact that the main purpose of weapons is to maim and kill. The words used to talk about weapons aestheticize them and their use. The use and effects of weapons are described with euphemisms, a metaphorical language disguising the painful and messy reality of actual wounds.[45] Shooting at animals, as described in *Jägaren* is in this way often euphemized as "greeting" them. "Even though the bird seemed to be out of range I saluted it," one hunter recalls, and with surprise observes the bird falling into the sea.[46] Sometimes the actual death of the animal is surrounded by greetings going both ways: "The Fox greeted me with its tail and disappeared," one hunter reveals after he shot at the animal, and while he is fumbling for more cartridges, the fox was again "greeted by a third gunshot by my youngest brother."[47] The relationship between the fox and the hunters, which ultimately ends with the death of the animal, is thus designated as an exchange of salutations. Beside being an example of ascribing agency to the animal, the kill is here rewritten as a more or less friendly encounter between human and animal.

In *Jägaren* the description of the actual killing or maiming of the animal is sometimes present but in these cases highly medicalized and "clean." The anatomy of the animal is referenced in the story, giving the impression of a hunter who shoots the animal with knowledge and precision. Thus in a dramatic recollection of a bear hunt, the hunter describes how his friend manages to fire a shot into the bear's shoulder. With anatomical precision the narrator tells how the bullet "passed through the lungs and damaged the spinal cord, forcing the bear to sit, after which he was put to death with a few rifle and revolver shots."[48] When the difficulties of shooting seals is addressed, the narrator describes how hard it is to kill the animals in the water. When sprayed with gunshots while resting on a rock, wounded seals roll into the sea, except one, "as he was hit in the spine, just below the neck" and therefore was unable to move.[49] In another account of a bear hunt, we are given precise and detailed information about where and how fired bullets hit the bear.[50] Often the animals were examined after the kill, and the stories relay how the bullets hit or what killed the animal.[51] According to Bourke the science concerning weapons and their use, from ballistics to medical analysis of gunshot wounds, is abstracted and described in clinical terms rather than addressing the individual casualties, a tendency Bourke describes as the use of a "numbing techno-speak."[52] This tendency is also apparent in hunting stories.

Thus the taboo relationship with weapons and the industry that produces them can fruitfully be compared with the way the killing of animals has been discussed within a Western, and in this case Swedish, context. The same distancing, abstracting language is ubiquitous when the actual act of killing animals is discussed.

NEGOTIATING ETHICS, ECOLOGY, AND ANIMAL AGENCY

The hunting stories in *Jägaren* reinforce the negotiation of ethics where both technology and ideas about violence and cruelty are involved. The ethics of the stories are, however, not necessarily obvious to the modern reader. The tales are often gruesome and little concern is given to the cruelty involved in the hunt, the maiming and stressing of animals. Yet ethics are embedded in the stories. Within the sometimes comical and pleasurable tales, ethics are continuously discussed. Technology is used to facilitate the mediation of the use of violence against animals. The interpretation could be that technology enables the exploitation of animals, while maintaining an ethical framework that makes the mistreatment of animals bearable.[53] The different positions seem to rub against each other: on the one hand, the anti-cruelty regime, with its demands for the welfare of animals; on the other, the gentleman hunter with his status as a knowledgeable scientist and amused sportsman. Their incompatibility is a point of departure for Swedish twentieth-century human-animal relationships. The compromise is accomplished by using several methods that characterize the wider human-animal relationships in Sweden. As we have seen in other chapters of this book, technology is used as a mediator, a facilitator to "communicate" with animals. This is what is represented in the *Jägaren* hunting stories: technology is used as a bridge between human and animal. The fact that the purpose of the bridge is to enable the human to kill the animal is disguised in euphemistic language. Ethics, ecology, and animal agency are constantly present in the texts, however covertly, being negotiated in an indirect way. The stressing of a relationship between hunter and animal, by describing the hunt as a game or a dance, is connected to the ecological ideas that were emerging at the time. Ecological thinking, the idea of animals existing in a web of nature, being part of a landscape, caused hunters to question the way animals were killed. While encouraging the hunter to become knowledgeable about the killed animals, the stories simultaneously promoted knowledge about ecology and animals as important inhabitants of nature, with lives of their own.

The stories in *Jägaren* are structured to emphasize the hunter as a person who enjoys being in nature, who appreciates animals, and who relishes the surrounding landscape. The never-absent weapons, however, make sure that the reader is always aware that the endeavor has a specific purpose: to kill animals. The technology appearing in the stories nevertheless ensures that this killing is done in an acceptable way. In the stories the knowledgeable, upper-class hunter suggests that the killing is performed responsibly; agency is distributed in the narratives; the chronology of the tales is structured in order to emphasize the ethics of hunting; and animals are explained as being killed by technology, rather than by humans.[54]

The hierarchical outline, where the upper-class hunter visits the rural and backward environment where the hunting takes place, enforces the view of the huntsman as superior in class, in scientific understanding, as well as in morals. This unequal relationship embedded in the stories enables the hunter to digress from strict ethical rules (such as using a sufficient caliber of ammunition to swiftly kill the animal) while maintaining the position of being an upholder of moral fair play. The temporality of the stories functions in a similar way, giving the impression of the hunter as meticulous, fearless, and knowledgeable.

The relating of sound in the stories also underlines the tale as a hunting story. The sounds of gunshots and barking dogs are loaded with meaning, which are strictly understood from the hunter's perspective. In spite of the animal often being awarded agency, this does not assert itself in relation to sound. The animal agency in hunting stories is clearly restricted, and is most obvious when it is described as being a part of the game, as voluntarily entering a dance with the hunter.

The rationality and modernity of technology used seems to guarantee that the death of the animal is not messy, particularly painful, or nasty in any way. In *Jägaren*'s emphasis on modern guns, its occupation with technological aspects of the hunt, as well as the ascribing of certain forms of agency to the animal all point to an ambivalent position in relation to the killing of animals. The practice of using technology to obscure the exploitation of animals is ubiquitous in Western societies. With the emergence of animal welfare movements, animal abuse was framed as outdated, in law and practice. Thus in a wider discourse on human-animal relations, technology functions both through rationalizing the killing of animals and obscuring the corporeal "messiness" of the procedures.

The remorseful hunter and editor of the journal, Hugo Samzelius, permanently avoided hunting for sport after he quit editing the journal. Yet he remained a professional hunter for wildlife management and was appointed as county forester in Stockholm in 1910. His interest in dogs persisted, and he was from that same year secretary of the Swedish Kennel Club and edited its journal.[55]

CHAPTER 5

DOGS WITH ANTENNAS AND THE COPRODUCTION OF HUNTING IN A GPS-ENABLED WORLD

FINN ARNE JØRGENSEN

On a chilly September morning in 2018, I found myself among a group of hunters in the Swedish countryside. I had gotten a ride from my hotel in downtown Umeå at 4:45 in the morning with two hunters, both professors at the Swedish University of Agricultural Sciences (SLU). We drove up the E12 toward Lycksele, then along the 363, and then on increasingly smaller roads, eventually following tiny forest roads built and maintained by timber companies for access to their vast tree plantations. These roads are not found on Google Maps and are used by the hunters with the blessing of the lumber companies, who depend on hunting to manage the moose population that will damage their tree plantations. Not long after 6 a.m., ten cars converged on an intersection of these roads. The hunters that emerged from the cars were here to take part in an age-old Scandinavian tradition: the annual moose hunt.

The hunters were a fairly varied group, as hunting teams go. Most were older men, though there were three women and several younger people. Most, if not all, were affiliated with SLU; the university's hunting team had been active since SLU opened its Umeå campus in the late 1970s. Some of the members had been part of the hunting team since the beginning; they knew the landscape they were about to enter like the back of their own hand. Others had joined later on and were still learning the ropes of hunting—the practical skills, the spoken and unspoken social codes, and the lay of the land. Most were Swedish, but some of the younger members of the group were international. My place in this group was definitely as an outsider—for one, I didn't carry a gun. I am no hunter and have never shot at an animal. However, the particular composition of the hunting team as researchers and professors meant that they understood and accepted what I was trying to do. I was given to understand that this would not necessarily be the case with more traditional hunting teams.

Why then did I find myself here? My goal in joining this group was to observe how hunters and hunting dogs interacted during the moose hunt, and then in particular to understand which role locative technologies like GPS units played in this interaction. I had researched this topic for a while and wanted to see for myself how it all worked in the field. My direct motivation for this fieldwork, though, stemmed from a simple observation made from reading hunting magazines. I had done a literature review of these magazines from the late 1990s, and noticed that in the mid-2000s something changed. Suddenly more and more dogs had antennas in the photos in the articles, much more often than the written content in the magazines indicated. It showed that hunters rapidly and extensively adopted GPS collars as a common part of the hunt. This made me curious: How did this change the way the hunters communicated with their dogs and how they sensed and made sense of the landscape?

The hunting team had brought several dogs: one was milling about, clearly excited, as the hunters got ready; the rest were in their cages in the cars. Each of the dogs was equipped with a large, bright orange GPS collar with an antenna, as is the case with most hunting dogs today. Their movement through the Swedish forest landscape this day would be synced with satellites moving in orbit so that the hunters could follow the dogs on the screens of their handheld GPS units. This scene is illustrative of the practice of hunting today, which integrates new technologies and old traditions in ways that reveal much about human relationships with animals and nature.

In this chapter I explore how Swedish hunters have used GPS and radio trackers as tools to enable and reshape deep spatial relationships between humans and animals. Hunting has always relied on tools that are intrinsic to its practice. These tools enable action, particularly killing at a distance (with spears, arrows, guns, and other weapons), as well as the exchange of information between hunters, prey, and domesticated hunting animals such as dogs. These hunting animals are a kind of mediator, translating and transmitting the relations of landscape and prey for their more spatially-bound operators, who use the data they get from their dogs as part of a mental visualization of the movements of prey. Since the 1970s a new set of geolocative technologies became available on the marketplace. Tracking collars augmented the remote sensing of hunters through dogs and thus further deepened the technologically mediated relationship between hunters and their dogs. Hunters have increasingly integrated media tools such as GPS transmitters into their practices, extending, enriching, and quantifying these predatory relations. It was these relationships I was interested in observing through my fieldwork.

This chapter examines the integration of informational media, digital communication, and locative technologies into moose hunting practices, deepening the study of technology in hunting through analyzing the circulating relationship between this particular set of digital technologies and the knowledge and practice of hunting. Taking as its starting point Swedish hunters' long tradition of actively reflecting on their own practices, the chapter will examine what kinds of discussions the hunters had about the introduction of GPS tracking and other communication technologies, as well as how they have been implemented in the field. I explore how such devices find their way into—and how we might use them to understand—what we could call multisensory mediated landscapes of hunting. What were the consequences for the way hunters know and value nature, negotiating between "tradition" and "authenticity"? What does GPS mean for the relationship between hunter, dog, and prey?

INTO THE FIELD

Hunting has long traditions as a field of study within the humanities and social sciences. Historical overview works tend to be either chronologies of hunting from the earliest times or demonstrations of the cultural value of hunting.[1] Matt Cartmill's award-winning *A View to a Death in the Morning* discusses hunting as one of the origins of humanity, and traces hunting traditions through the present.[2] Internationally, research on hunting and wildlife management has gone through a reorientation the last few years. The history of hunting provides us with a key insight into general public attitudes to nature and environmental change. For instance, many have examined how hunters develop strong social norms around their practices.[3] They were pioneers in animal conservation in the late 1800s, largely tied to a wish to be able to continue hunting certain species—and to a lesser extent guilt over endangering species.[4] Contemporary hunters also have to practice hunting in ways that are acceptable to the public.[5] Many debates over the ethics of hunting are concerned with the use of modern technologies such as ATVs, GPS trackers, drones, camera traps, and so on, in ways that conflict with ideas of traditional hunting.

Hunting comes in many guises, from subsistence, sport, canned, and trophy hunting.[6] While there is an element of subsistence to Nordic moose hunting, it is mostly a recreational activity. Hunting is an age-old tradition in Scandinavia, as in most other places, but it has been a long time since hunting for subsistence was a necessary activity. In the period this article covers, from the 1970s, hunting has been predominantly a leisure activity. At the same time for many people it is also a way of connecting with the history of their family or the local landscape, or engaging in local traditions. Through these technologically mediated connections, hunting enables particular relationships with nature. Hunting "may be regarded as a form of behavior that unites men with nature rather than alienating them," wrote philosopher Paul Shepherd.[7] This obser-

FIGURE 5.1. Dog with antenna. Photograph by Finn Arne Jørgensen, 2018.

vation is mirrored in the reflections written by hunters, such as the Norwegian hunter and journalist Torolf Kroglund, who stated that hunting takes people from being an *observer* to being a *participant* in nature.[8]

Lyndel V. Prott has discussed hunting as intangible cultural heritage, arguing that it is a tradition that connects the present with the earliest days of humanity.[9] The Forum feature of the journal *Environmental History* reflects on the many different ways in which the story of wildlife in North America had been told. These were "tales of ecological decline, tragedies of the commons, chronicles of scientific discovery, parables of ethical redemption, clashes of values, reorganizations of spatial relationships, expansions of federal authority, and struggles of conservation versus social justice."[10] Hunting with guns or cameras is a fundamental way of engaging with nature in general and wildlife in particular. In general, recent environmental histories of wildlife management have focused on governmental agencies, scientists, or environmentalist action groups, rather than on hunters themselves.[11]

The most common dogs used in moose hunting are Norwegian/Swedish elkhound (*norsk elghund/Jämthund*), Russian-European Laika, or Karelian bear dog.[12] The Norwegian elkhound is a modern breed produced in the 1800s specifically to hunt elk (i.e., moose).[13] Dogs are a critical part of moose hunting—and as important a part of the hunting team as any of the people. Their place in hunting was a key part of the domestication of dogs throughout history, and hunting dogs are still quite common in Scandinavia. Hunting dogs are also technologies, bred with very particular abilities over time, as argued by environmental historian Edmund Russell.[14] This refers both to their breeding and their equipment: while hunting, they wear reflector vests, just like us, to be visible. And they have antennas, as we see in figure 5.1.

Technology has an important place in the story of wildlife management, but this story has generally been told in limited ways. Etienne Benson has examined the role of wildlife tracking technologies in wildlife biology and management but did not look into the parallel story of hunters and their use of locative technologies.[15] In general, histories of hunting tend to pay much attention to weapons and their development from the dawn of time to the present, but spend far less time on exploring the role of other technologies. The general descriptions of technology to a large degree align with what historians of technology classify as internalist studies: chronological stories of changes in technical details with little regard for the interaction between technology, culture, and society at large.[16] Such histories are necessarily limited in their understanding of change and continuity in hunting.

The history of locative technologies in hunting demonstrates the need for understanding such technologies in the context of their deployment and use over time, with a keen eye for how they enable new forms of co-construction of space. A number of historical studies of GPS have been published in recent years, some by professional historians, others by engineers. "GPS has become an invisible piece of infrastructure," writes historian Paul E. Ceruzzi.[17] It is our task to make GPS visible again, to reveal its connections with the world in which it operates.

From its origins as a Cold War military technology, GPS has found far more commercial and nonmilitary uses than its creators envisioned. Through one-way communication with a network of around twenty satellites orbiting Earth, receivers can determine their location with a high degree of precision. When President Bill Clinton turned off signal scrambling for nonmilitary users in 2000, GPS receivers suddenly became much more accurate.[18] These receivers also got much smaller over time, opening up a whole new range of applications. Starting around the time of the GPS descrambling in 2000, more and more handheld GPS units started appearing in hunting magazines like *Svensk Jakt*. Such handheld devices were a primary nonmilitary use of GPS technologies. Eventually dog collars appeared in the magazines, enabling the linking of dogs and hunters through GPS.

Technological tracking and communication in hunting did not start with GPS, though. In fact, the GPS that hunters use today is an extension of radiotelemetry, an older practice using radio transmitters and receivers to judge distance, as discussed in Etienne Benson's book *Wired Wilderness*.[19] It is unclear exactly when Scandinavian hunters figured out that they could use radio-tracking collars to keep track of their hunting dogs and not just wild animals, but the Finnish company Tracker started selling them as early as 1977. As a technology the radio trackers were relatively simple, and could only point the hunters toward the dogs within a certain range. A tracker device used radio signals to indicate the direction of the dog, as well as the distance. It is unclear what the range was for these early models, but later ones could function up to ten km.

Looking at the pages of *Svensk Jakt*, the largest hunting magazine in Sweden published since 1863 by the Swedish Hunting Association, we find two types of stories about hunting dogs and GPS. The first is one of hunters eagerly embracing new technology. The second one, not quite so pronounced, concerns hunters trying to find out where the new geospatial tracking technologies that they so eagerly adopted fit into their established traditions and practices.

Browsing the pages of *Svensk Jakt* indicates that Swedish hunters adopted GPS tracking remarkably fast. When the Garmin Astro dog tracking system arrived on the Swedish market in 2005, it dramatically changed the market for dog tracking. The old radio trackers disappeared, and the advertisements in the magazines changed accordingly. After 2007 GPS-related products appear in almost all issues of *Svensk Jakt* in the New Equipment feature. In addition, images of GPS technologies in use appear in the photographs published with hunting stories that narrate and analyze a particular hunt. Even when these articles don't explicitly mention GPS technologies as part of the setup, readers can observe that so many of the dogs in the pictures have antennas on their collars. This was not the case in 2000, but by 2007 it had become a standard feature in the magazine. This was a significant change from the days of radio tracking, which was typically not shown in photographs. After the GPS breakthrough, however, it was a common scene for dogs to have antennas in general illustrations. GPS trackers had become an indispensable part of the moose hunt.

THE HUNT

The Swedish moose hunt takes place every fall, starting in early September in the north and in mid-October farther south. The Eurasian moose (*Alces alces*) is called the King of the Scandinavian forest, and can weigh up to 475 kg for males and 375 kg for females. It is a distinct and separate species from the American moose (*Alces americanus*). Scandinavians have hunted moose for centuries. During the 1700s the Fennoscandian moose population collapsed due to overhunting and pressure from a large population of wolves. In the late 1800s the moose population began growing again, in part due to new hunting laws, but also to the in-migration of moose from farther east. Today, the Swedish moose population has grown very large, with a total population of 340,000 and about 83,000 moose killed every year.[20] The numbers fluctuate somewhat, with a high point since the mid-1990s of 103,000 moose shot in 2003–2004.[21] Each hunting team has a quota that they can shoot every season.

The team that I followed practiced a traditional team-based style of moose hunting, using freely roaming dogs. In this style teams of hunters have specialized roles. Some hunters are placed with rifles on stationary posts. Others follow the dogs, whose role is to locate the moose. The dogs are bred to bark to signal the position of moose and play a critical role in this hunting technique. When I joined the hunting team in northern Sweden, they started with a morning meeting by the cars where they planned the day, discussed the landscape, and assigned people to posts. They all had maps that clearly showed the borders of the hunting area limiting where they could hunt. The posts are important—people must stick to their post and know where the others are. It is the responsibility of the team leader to assign people to their posts, firmly establishing safe directions to shoot. The hunters I spent time with during my fieldwork repeatedly stressed their competency in this area as better than most other hunters. There are shooting accidents in Sweden every year, and they were proud that they hadn't had any. Each hunter was equipped with a radio so that they could keep in contact with other hunters.

Hunting practices are deeply embedded in larger structures and traditions, such as the allotment of hunting rights, and of historical transitions from hunting as a sustenance strategy to a leisure activity. In Sweden hunters need a permit to hunt in particular places, and these are given out to landowners and often managed by the hunting teams. This means that as part of such a team, you have a particular plot of land that you can hunt in as long as you are on the team. It follows that hunters get to know this landscape rather well. While it can be argued that Swedish hunters generally hunt in local areas that they know rather well, it could also be that locative technologies like GPS enable hunters to navigate landscapes better than they could otherwise. The many articles discussing and reviewing GPS units in hunting magazines make it clear that hunters do not know the landscape as intimately as they might think. For instance, in a 2007 test of car GPS units, the envisaged use scenario was one of finding your way on small country roads so that the rest of the hunting team didn't have to wait for you. These roads generally belong to forestry companies and can definitely be confusing to navigate.

SENSING DOGS AT A DISTANCE

During the first session of my hunting fieldwork, I joined two men and two dogs to find one post some distance away from our starting point. Upon arrival at our designated post, we left one hunter there, armed with a rifle—he was to stand there in silence and look and listen for the dogs and moose. He considered himself relatively old-school; he told me that he saw the point of using GPS collars on dogs, but did not have his own GPS unit to track them. Many high-end GPS

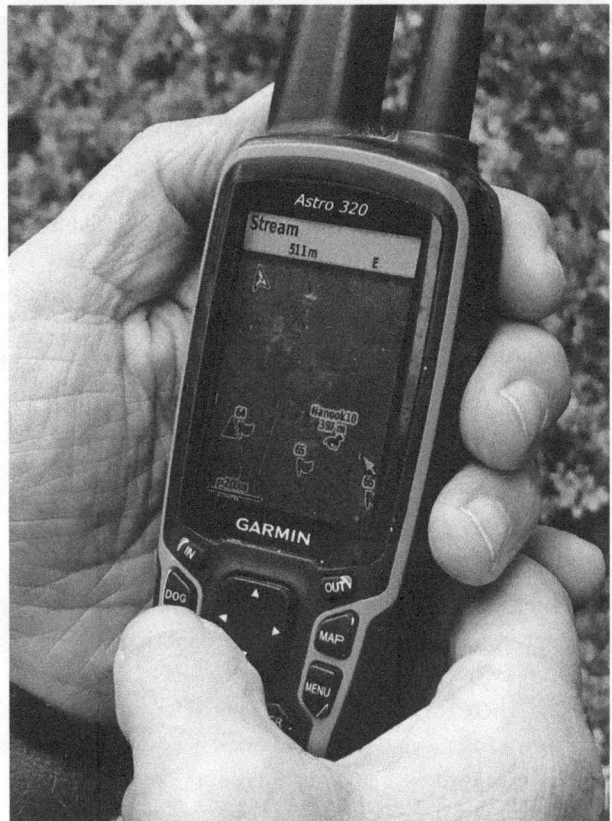

FIGURE 5.2. Following Nanook the dog on a handheld GPS screen. Photograph by Finn Arne Jørgensen, 2018.

units used in hunting can track the location of other users and dogs. He had a walkie-talkie, though, as everyone else did, so he would get updates on what was happening elsewhere.

I then walked off with the other hunter and the two dogs to look for moose. My guide, a Norwegian veterinarian, owned the dogs, one older and experienced and one younger who was still learning. As he explained it, hunting is not a skill or a behavior that can be trained, like you train a puppy to sit and fetch and walk—it is something much more instinctive. Ultimately, you have to just let the dogs loose and trust them to figure it out, typically with help from other dogs. He was a very experienced hunter who came here every year for the hunt with this team, driving up from his university in southern Norway. Not coincidentally, he held a guest professorship at SLU in Umeå and had a long collaboration with one of the professors here.

We released the dogs and followed them on foot. They were eager and seemed happy to be able to run free. Occasionally we could see them run by, but mostly they were out of sight. We then followed them on GPS, walking in the general direction they were heading, though much slower—the dogs are fast! The hunters know the landscape very well after hunting here for many years, so I could see how the hunter I followed connected the mental landscape, the physical landscape, and the digital landscape on the screen. This was no passive and virtual experience, but one full of active interpretation, as he explained to me what the dogs were doing and where they were going, building on his experience with the landscape.

After twenty minutes or so, we heard barking. The hunters talked on the walkie-talkie—someone had seen a moose, possibly two. In their discussions, they visualized where the dogs were moving and inferred where the moose had to be from the dogs' position. It was strangely intense, as the moose moved through the hunting grounds, but avoided the posts, almost as if the moose knew exactly where the people were. No one actually saw the moose after that first glimpse—they could only indirectly sense it through hearing the dogs and seeing them on the screen. Then the moose started moving toward the border of the hunting ground, about to go into another team's area. If this were to happen, the day's hunt would end for the dogs' owner, since they can't simply move into another hunting area. They would need to establish contact with the team hunting in this area to make sure everyone on the other team knew that people were coming and where they were. This is about safety, which means the tensions surrounding a dog moving toward a border get even more intense. What would happen in the next few minutes could ruin the entire day.

But then, we heard a shot. The team had one person on a post at the edge of the hunting grounds. He had come late, as he had forgotten a piece of his rifle. He had just gotten set up at his post, and then the moose came over the road. As he described it to us later, he shot, and the moose ran into the forest again. The dogs followed and when they got quiet, the hunters knew—could sense at a distance—that the moose had stopped, dead or dying. It was time for us to find it.

We walked back to the car and drove along the small forest roads to the post where the moose had been shot. In the forest, not far from the road, we found the moose, dead. Both dogs were nearby—the older dog had "claimed" the moose as hers, and the younger dog was just a little bit farther away. I had seen many live moose over the years, both wild and in enclosures, but never a dead one. It was still very warm, after running for quite a while. While this was a young bull, it was very big. They smell a bit like sweat, not entirely unpleasant but certainly noticeable. I had to help turn the moose over so that they could do some quick field butchering to gut it and take out the stomach, as it could ruin the meat if its contents leaked out.

Observing the hunting team in the field made it obvious how moose hunting with dogs is not a solo activity. The smallest team can consist of just a person and a dog, which makes interspecies cooperation a critical component of hunting. The division of skills and tasks among the human and nonhuman participants on the team is important, as are the communication and coordination among the team members.

There can be no doubt that in this hunting process, various technologies extend the senses, like binoculars. Communication is absolutely critical for this sensing, as hunters need to coordinate a large mental map of the area, including themselves, the dogs, their teammates, and the moose, keeping landscape features and boundaries in mind. Most essentially, they need to know what their dogs are doing. In the early 2000s, when dog tracking with GPS was still fairly new, *Svensk Jakt* featured a story about two hunters following their dogs on the screen. They watched the dogs slowly follow the scent of a bear through the forest, over roads and streams. After about an hour, a few kilometers away, they suddenly sped up—"the bear is probably trying to run away," the hunters interpreted the data. "They tend to do it before they take a stand," one of them whispered (as if the bear could hear him through the data connection).[22] This is an interesting form of sensing from a distance, of reading the signals and movements of the dogs—which they did based on sound earlier, but now on a screen. Where the dogs are is one thing, but what they are doing is another, requiring experience and interpretation to determine.

When hunting moose with a dog, the hunter and dog build a sense of trust and communication at a distance. The dog runs free to find the quarry and needs to signal the hunter that it has found something. If it can't find anything, the dog needs to come back to the hunter. As such, dog and hunter need to be aware of each other's location in the landscape. This is why many hunting dogs bark so much—the hunter can not only hear where the dog is and what it is doing, but, with some experience, can also judge the distance.

But listening to barks is demanding and also uncertain. If the dog stops barking, it's hard to find. If the dogs follow the moose outside their designated hunting area or just roam too far, you risk losing them altogether. This is not like losing another piece of equipment, as there is often a tight bond between hunter and dog. While one can consider the dog a technology for hunting, they are never *only* a technology. The relationship between owner and dog is often close, and the dogs also represent a considerable investment in terms of time and training, so the risk of loss is a strong motivator for dog tracking. The pages of *Svensk Jakt* are full of articles about what happens if dogs don't come back. They might have gotten distracted, lost, attacked by bears, wolves, or moose, or just be poorly trained. Hunters like to tell stories of tragedy and redemption, of searching for lost dogs for four days in −30°C in the winter, and then miraculously finding the dogs alive, to much joy for both dog and owner.

Dog tracking also provides some insurance against other and more unwanted forms of animal encounters. Hunters consider wolves a constant threat in Sweden, even though the national wolf population is only around three hundred individuals at present, mostly roaming freely in the south-central parts of Sweden.[23] Wolves will kill dogs they come across, so if hunters see any wolf signs, they need to find their dogs as soon as possible. There was an expected rise in attacks from the growing wolf population from the 2010s. This is fascinating in hindsight, since there aren't all that many wolves in Sweden; they were declared extirpated from the country in 1980, but repopulated from Finland and Russia. But still, there have been fifteen to twenty-five reports of dogs killed by wolves every year. *Svensk Jakt* features many articles about such incidents, with quite gruesome pictures. Such stories express anxiety about location in space that hunters and dogs share with other beings. For hunters GPS tracking has therefore become an important safety net.

THE COPRODUCTION OF HUNTING IN A GPS-ENABLED WORLD

Following the successful kill of the moose bull, the team gathered nearby to load it onto a sled that was pulled by an ATV up to a trailer on the nearby forest road. Then the hunters started a campfire and got out their lunches. While drinking coffee and roasting sausages over the fire, the hunters discussed the morning session and the story of the kill was retold many times. Afterward the maps came out again and people were reassigned to new posts. I got to stand on a post with another hunter, so that I would also get to observe this aspect of hunting.

Standing on a post is of course much more static than following a dog around, but requires no less active sensing. I was safely positioned behind the hunter with his gun, but I was scanning the landscape for movement in just the same way as he did. After an hour or so, one of the hunting dogs and his owner passed by, but we saw no signs of moose. We listened intensely to the sounds of the forest, amplified and slightly distorted by the safety ear muffs we wore to protect our hearing in case of gunshots. These let through most sounds, but would clip excessively loud ones. We heard one bark in the distance and the occasional rustling and beating of wings as nearby birds—grouse and ptarmigan—were startled by something. But there were no more signs of moose. In the end we called it a day and met up again by a junction of the forest roads for a final debriefing. The other hunters would go out again several more times during the few weeks in September of hunting season, but I would finish up my Sweden visit and go home to Norway.

Hunters tell stories to express values and concerns. Since the introduction of radio-tracking dog collars in the 1970s, Swedish hunters have debated and questioned the increasing dependence on hunting technology. As radio tracking became more common, they worried certain skills and traditions that they valued would disappear. Some of their concerns were more direct than others. For instance, in one *Svensk Jakt* article they warned, bizarrely, against collars with antennas since "they could poke out people's eyes," making the dog owner responsible.

In most cases, though, hunters engaged in careful integration of new technology into existing practices. Many hunters would argue that GPS has not fundamentally changed the way they hunt—for them, it is a tool to use when training dogs, and should only be used to make sure the dog doesn't do anything stupid. The danger, as one hunter phrased it, was that uncritical use of GPS tracking resulted in dogs that ranged too widely, too far away from the hunter. Some stated that it would be dangerous to let this technology take the place of the contact between hunter and dog. The dog must be trained to return to the hunter when it's not chasing prey; the ideal was that the dog should seek the owner, not the other way around. In other words, hunters could not let the dogs know that they know where the dogs are. One example of this from a *Svensk Jakt* article tells of an experienced hunter and dog trainer using the tracker to go find an old and well-trained dog. After this incident, though, she no longer listened to his signals or returned whenever she wore the tracking collar; she would instead wait for him to come get her with his warm car. And the fascinating thing is that she would return on her own when wearing just a regular collar. As the story went, the dog could tell the difference between the different collars and decided on her own how to act.

One can of course wonder about the validity of these hunting stories, but I think they are indicative of cultural negotiations about the relationship between dogs and hunters and what roles technologies play in these relationships. What such stories also tell us is that dogs have agency.[24] Hunting requires interaction between human, dog, and moose (and other species!). This interaction is placed within a landscape. Locative technologies have augmented the practices of sensing and have brought that landscape into a digital model featuring both dogs and humans—but not the moose, as most hunters would consider that to be cheating. "The moose should get a chance. If the hunters don't pay attention, they only have themselves to blame." "Sometimes it feels like the dog guides [*hundförerna*] can't take a step without reporting where they are. But the worst is when they follow the dogs on their GPS and tell the hunters with the guns where the moose will come."[25] Dogs have to be trained to find their place in such a system—to do what is expected of them. It is not simply a question of the dog being a tool that GPS allows you to control, however. They are also active users of technology, and thus take part in coproducing both the technologies and the landscapes experienced through them. Dogs are not simple tools that can be trained for a particular goal. They have agency and can be working toward a shared goal, but not necessarily so. It is a matter of communication and trust, and hunters have had to work hard to integrate technology like GPS into this communication. That is another thing that didn't happen automatically.

Today, hunters have embraced GPS and digital maps. There are web services where they can customize maps for their own hunting teams and that can create hunting logs with pictures and everything. In more recent issues of the hunting magazines, GPS shows up in almost every hunting narrative (they have a few in every issue), but they don't make a big point of it. It has become as much a part of the hunt as the gun. GPS logs have become important for evaluating the hunt afterward. The geographical data can be displayed on top of a map, showing where the dogs and the hunters have gone. The Swedish Hunting Association developed its own smartphone app, WeHunt, to exploit the possibilities that such technologies provide. This app allows you to buy the official high-quality terrain maps with details about property rights, buildings, water bodies, roads, topographic information, weather forecasts, sunrise and sunset times, and so on. Using this app, hunters can keep track of and chat with the rest of the hunting team in real time, and the app will look after the hunting area boundaries, as well as display which animals are legal to hunt. Recent versions even have an augmented reality function, where hunters can hold up their phones and see the camera view of the landscape around them with the aforementioned data about dogs, fellow hunters, and landmarks as points on the screen, showing where they would be in the landscape.

Hunting is a technological system, with a number of both human and non-human integrated components creating a shared space. Technology provides a connecting bridge between the different timescales of human and environmental change. However, technology is not a neutral mediator. The particular constellation of technologies and users in use at any given point in time shapes our understanding of both the mediated phenomena and ourselves. In the case of GPS, we delegate both sensing and decision-making to a particular technology, placing a certain trust in the vast technological infrastructures that make the little screen in our hands display directions to where we are going. In the process of delegation technologies extend the human body, augmenting our experience of the world.

Hunting is an intensely spatial relationship between hunter, dog, prey, and landscape, one that engages senses and skills. These are acquired through training, socialization, and breeding. In the learning and the doing of hunting, people are enacting traditions. Pervasive geolocative technologies like GPS have both qualitatively and quantitatively changed the hunt, the hunter's experience of hunting and nature, as well as its management. In doing so GPS has reconfigured the spatial human-animal relationship. Hunters make sense of space and place in processes that are constantly changing, relational, and tied to social values. We must include the digital layers of information that today cover the landscapes around us. Hunting is a cultural heritage, one that, like any other, is being reinvented for the digital age. Hunters love their tools—and hunting is infused with technology, but with constant negotiations about the appropriate technology. I think we are at a junction right now where hunters are changing the way they interface with nature (but probably not for the first time). As Danell and colleagues observe, "Cables, antennas, and information via GPS receivers and mobile phones today gives less room for the magical. At the same time the modern tools give a new dimension to the hunt. When one now can follow the movements of the dog through the landscape in a better way, it opens up for interpretation and perhaps new forms of speculation."[26]

This chapter has shown how the relationship between hunter, hunting dog, and prey changed when hunters adopted GPS devices in their hunting practices. The increasing use of such technologies by hunters reveals deeply embedded ideas of embodied traditions of learning, navigating, and sensing a landscape. The hunters' negotiations over how to best and most appropriately integrate new GPS technologies into their established traditions demonstrate that few hunters think of their own use of GPS as inappropriate, but that they equally often think that others use the technology in ways that threaten age-old traditions.

CHAPTER 6

WOLF RESEARCH AS MULTISPECIES KNOWLEDGE PRODUCTION IN FINLAND

HETA LÄHDESMÄKI

During the 1990s the first wolf individuals were radio-collared in Finland. Since then, radio-collared wolves have provided information on the whereabouts and behavior of Finnish wolf packs and on the species as a whole. In this chapter I focus on scientific wolf research conducted with radio and GPS telemetry in Finland from 1998 to 2004. I look into material related to wolf research as well as newspaper and magazine articles to find out if and how wolves and humans produced knowledge together.

Knowledge can be understood as reaffirming the human-animal divide. For instance, John Berger has stated that the more we know about animals, the further away they are: "What we know about them is an index of our power, and thus an index of what separates us from them."[1] In a theoretical sense, this is true. However, I argue that knowledge production can make humans and nonhuman animals closer in a physical sense. In this chapter wolf research conducted with the help of radio and GPS telemetry is seen as a spatial, material, and technological entanglement between wolves and humans. It is an example of *natureculture*, a phenomenon in which creatures, like wolves and humans, and things, such as technological gadgets, categorized as human or nonhuman, nature or culture, start to entangle, as science and technology studies scholar Donna Haraway has written. Humans and nonhumans, nature and culture are entwined.[2]

Knowledge production and nonhuman agency have been explored, for instance, by actor-network-theory scholars Bruno Latour, John Law, and Michel Callon.[3] Feminist science studies scholars Linda Birke, Mette Bryld, and Nina Lykke have also looked into the processes of constructing knowledge, using the notion of performativity, in the sense that the feminist theorist Karen Barad has developed, to analyze how laboratory rats participate in the creation of knowledge alongside human scientists.[4] By being inspired by Barad's views on performativity and agency and the way Birke, Bryld, and Lykke work with them, I seek to show that individual wolves took part in the process of creating knowledge while carrying a radio or GPS collar. I show how technology increased human knowledge of wolves but also made humans and wolves interact closely. Even though the human need to know gave rise to the use of telemetry, the wolves and the technology often dictated the research. I also attempt to view wolf research from the wolves' perspective and show what this intra-action meant for the individual wolves in question. I aim to critically examine the benefits and disadvantages of radio and GPS telemetry and argue that wolves benefited from the research, but that it also raises ethical questions, like all forms of multispecies coexistence.

WOLF RESEARCH TECHNOLOGIZED

Field research on wolves began in Finland in the 1960s.[5] For a long time it relied heavily on information gathered with the help of volunteers observing wolves and their tracks in the snow.[6] Wolf research intensified in the 1990s with the introduction of radiotelemetry, though wolf researchers and conservation officials had already discussed the use of this technology in the 1980s. The Council for Natural Resources stated in a 1986 report that "in order to estimate the population sizes for large predators, it is vital to know the size of these animals' territories in Finland. . . . We should get more experience with radiotracking in Finland, even though comprehensive examinations would not be possible right away."[7] In March 1998 the Finnish Game Research Institute's researchers radio-collared five wolves in the Kainuu region in eastern Finland.[8] The project gathered a lot of media attention. According to the newspaper *Helsingin Sanomat* and the hunting magazine *Metsästäjä*, by monitoring the wolves' movements, the researchers wished to find out how many wolves there were in certain areas, as well as what their life history and the size of their territories were. This information would help estimate the number of wolves living in Finland.[9]

From 1998 to 2004 sixty wolves from eight territories in the regions of Kainuu and Northern Karelia were equipped with radio collars.[10] The number of tagged wolves in Finland was relatively modest: over a hundred radio-collared wolf packs had been studied in Alaska by 1991, and about a thousand wolves had been radio-collared by 2003 in Minnesota alone.[11] Finnish researchers used VHF (very high frequency) radio waves to track the wolves' whereabouts. From 2002 on the wolves were equipped with collars with both GPS and radio.[12]

Wolf research intensified in many countries during the second half of the twentieth century, and researchers started to use different kinds of technical equipment and machines. The introduction of radiotelemetry in the late 1960s was part of this intensification and technologization.[13] In the 1950s American wildlife biologists came up with the idea to use miniaturized radio tags and collars to keep track of individual animals, and incorporated Cold War–era surveillance technologies into their practices. Radio-tracking was an essential part of wildlife research in the United States by the 1980s.[14]

Similar to the Swedish moose hunters studied by Finn Arne Jørgensen in the previous chapter, wolf researchers adopted radio and GPS collars as a common part of their practice. As historian Etienne Benson explains, wildlife radio-tracking and radiotelemetry have had an enormous impact on the everyday practice of wildlife biologists.[15] Researchers needed a lot of equipment beside the collars: receiving systems for VHF wildlife telemetry comprised radio receivers, antennas, and cables to connect the antenna to the receiver, accessories such as headphones and chargers, counters and decoders, as well as recording devices. GPS and global system for mobile communications (GSM) receivers could also be used, and when that was the case, the researchers also relied on satellites.[16]

Wolf conservation benefited from the intensifying field research. Wolf conservation started in Finland in 1973 when the species was protected, outside of the reindeer herding area in the north that covers about 36 percent of the country's surface area. Protection became stricter in the 1990s when Finland became a member of the European Union. The Nature Directive demanded member countries gather more knowledge on animal species under conservation.[17] Ministry of Agriculture and Forestry officials made the decision on the number of wolves that could be killed annually without threatening the vitality of the wolf population, and those decisions needed to be based on scientific research. It was therefore essential to know more about the wolves' life history and territories as well as the size of the wolf population to protect the species successfully. Previous research methods had not provided all the information needed.[18] Head wolf researcher Ilpo Kojola said in *Metsästäjä* in 1998 that the researchers believed radiotelemetry could provide new information and help minimize uncertainties when estimating the population size.[19] As biologists Mark R. Fuller and Todd K. Fuller state, "Radio-telemetry can be used to gather information that is neither practical nor possible to obtain with other methods from rapid-moving, wide-ranging, and secretive carnivores."[20]

From the moment radiotelemetry was implemented in wolf research in Finland, it became part of the human-wolf relationship. Finns tried to control the wolf population and used their knowledge of wolves in the attempt. Some scholars interpret the gathering of knowledge as a procedure that widens the imagined gap between humans and animals. According to John Berger, "Animals are always the observed. The fact that they can observe us has lost all significance. They are the objects of our ever-extending knowledge."[21] Philosopher Lori Gruen and researcher Kari Weil also state that knowledge is something that emphasizes animal otherness; the other is something or someone that can be known and understood.[22]

Historian Jon T. Coleman has studied the shared past of wolves and humans in North America. He describes how modern wildlife biologists, with the help of various technical gadgets, spy on wolves who are clueless concerning their status as a closely observed and well-monitored subject.[23] Radio-collared wolf individuals were also closely observed in Finland, many for several years. Their movement and the everyday events of their lives were known by researchers and also the general public because of media interest.[24] The digitalization of the animal world has been criticized because of this transparency. As author and journalist Alexander Pschera writes, it is easy to argue that the animals tracked by people are no longer able to hide from humans. One can ask whether information on their locations should be suppressed in the case of endangered animals.[25] Various ethical questions are linked to the digitalization and technologization of human-animal relations, including in wolf research.

INTERSPECIES COLLABORATION

Coleman's notion of wolf research as spying is accurate, but one can also interpret wolf research in a different way. Nonhuman animals do not have to be understood only as *targets* or *objects of* research but also as *actors participating in* it. To do so Birke, Bryld, and Lykke have used the concept of *performativity* in the sense in which Barad has reformulated it. Judith Butler developed the concept of performativity as an approach to the feminist theorizing of queer perspectives.[26] Barad uses performativity to show how matter matters, that it is an active participant in the world's becoming, in its ongoing intra-activity.[27] According to her, "Practices of knowing cannot be fully claimed as human practices, not simply because we use nonhuman elements in our practices, but because knowing is a matter of part of the world making itself intelligible to another part.—We do not obtain knowledge by standing outside of the world; we know because 'we' are *of* the world."[28] Birke, Bryld, and Lykke look at laboratory rats and argue that they are participants in the creation of their own meaning and of the history of scientific knowledge.[29] Laboratory rats differ from wild wolves, but of course the concept of performativity helps us to see that wildlife research subjects are also active participants in the creation of research data. Wolves participated in the research by carrying collars and living their own lives. They did so without giving consent, and it seems that sometimes they carried their collars unwillingly, as I demonstrate later in this chapter. Still, I argue that one can interpret wolf research and tracking as interspecies intra-action or collaboration.

Radio-tracking brought humans and wolves together spatially and physically. The collaring of the wolves took place during winter. Finnish researchers and research assistants drove a snowmobile alongside individual wolves and used a neck-hold noose attached to a pole to capture them. In order for them to capture the wolves, certain snow conditions needed to be present: it had to be soft and at least 80 cm deep. The captured wolf was placed in a wooden box and kept there for half an hour before being injected with a mixture of medetomin and ketamine (with a dose ratio of 1:20). The researchers then attached a collar around the wolf's neck. Once the wolf was collared, it was marked with ear tags, weighed and measured, and hair and saliva samples were taken for further genetic analysis. The researchers also checked its teeth. After that the researchers placed the wolf back into the box and injected an antagonist drug (atipamezole, with a fourfold dose compared to the dose of medetomin). The wolves were released after recovering.[30]

The collaring brought wolves and humans into close bodily contact. Pschera describes tagging as a direct connection between humans and animals.[31] Birke, Bryld, and Lykke call the intra-action taking place in the laboratory between human scientists and rats a choreography, a "co-creation of rats and humans, through their daily intra-actions, to *produce* the practices of science." The rats and humans are "dancing to the tunes of experimental protocols."[32] Similarly, the behavior and biology of wolves dictated the collaring procedure. Wolves are strong runners, so they had to be exhausted in order to get caught. When rolled over, wolves submit, so it was quite easy to get them into the box. The drugs available also influenced the procedure: the anesthesia used could not be given to an animal under stress, and therefore the wolves were put into the box to calm down.[33]

When researchers and research assistants put collars on wolves, they were under anesthesia. This was because wolves—unlike rats—would not have danced to the tunes of the procedures needed for collaring, measuring, and sample collecting. Still, the researchers and assistants had to know how to get the still-awake wolves into the box, handle the anesthetized wolves, and so on. They also needed to be fit because the chasing part was rough, according to an interview of research expert Seppo Ronkainen on *Yle News*.[34]

After the wolves woke up and were released, they started to participate in the research more actively. Unlike the laboratory animals that Birke, Bryld, and Lykke write about, the wolves were studied in their own environment, living their own lives unguided by the researchers. This was, in fact, the purpose of collaring the wolves: the researchers wanted the wolves to behave "normally" and to be able to collect data on that.

The researchers' closeness to the wolves did not end with collaring. After the collared wolves were released, the researchers observed their movements from both afar and nearby. Like the Swedish hunters and their hunting dogs in the previous chapter, Finnish wolf researchers used technology to be aware of the wolves' whereabouts. They tracked the radio-collared wolves regularly throughout the year, two to five times every week, through ground-tracking.[35]

According to Finnish researchers, "Tracking was made by triangulation of at least two directional bearings recorded at known remote locations. Researchers used vehicles with car-top dipole antenna systems to locate the positions where VHF signals could be heard. After that, location estimates were obtained using a Telonics H-Adcock hand-held RA-2AK antenna with Telonics TR2 and TR5 receivers."[36] In July 1998 *Helsingin Sanomat* described the workday of two wolf researchers. The researchers climbed into a 40-m-high fire observation tower in Kuhmo, Kainuu, and pointed a radar antenna at every compass point. The receiver started to beep and showed that a male wolf called Ukri was moving somewhere in the south, about 40 km away from the tower. The researchers drove close to the wolf's estimated whereabouts and used a smaller radio receiver to track the wolf. The receiver disclosed that the wolf was very close, only a few hundred meters away. They also got a signal from a female wolf, Maija. After finding out in which direction the wolf Ukri had gone, the researchers drove in the opposite direction so that the wolf's precise whereabouts could be defined.[37]

Helsingin Sanomat published an article about a deeper physical closeness between a wolf and a human researcher in April 2004. Alexander Kopatz, a German biologist working for the Finnish Game Research Institute, had tracked a male wolf called Noppe every day since the summer of 2003. Kopatz slept in an RV and changed his sleeping place according to Noppe's movements.[38] According to Erkki Pulliainen, a well-known biologist, wolf expert, and conservationist, the researchers kept generally a few hundred meters' distance from the wolves.[39] Still, in Kopatz's and Noppe's case, the researcher followed the wolf day and night and stayed relatively close to the wolf even when sleeping.[40]

From 2002 on VHS-GPS technology was used in Finnish wolf tracking. It made tracking easier for researchers who could downloaded the data from the VHS-GPS collars through a cell phone (GMS) connection. The collars were programmed to collect locations six times a day,[41] and the GPS receivers estimated the wolves' locations within seconds of receiving signals from satellites.[42] The collars informed researchers only about the their movement and location, but that information helped the researchers to go and look for the wolves, study the areas they had been in, collect feces, and look for denning sites, litters, and carcasses, thereby gathering more information about them.[43] Collared wolves, their movements and actions, along with the technology affected the researchers' movements and behavior and dictated what kind of data was collected.

Wolves that wore collars during the years 1998–2004 gave researchers information on many things. For instance, they examined the direction and distance of wolf dispersal, and the fate of the dispersing young wolves.[44] Researchers were able to find out where wolves dispersed and established new territories, which were detected by visual observations of collared wolves during aerial surveys and by following the movements of GPS-collared wolves. The boundaries of the new territories were defined through radio- or GPS-tracking.[45]

Telemetry research showed that the direction of the dispersal influenced the fate of the dispersing wolves. In northern Finland, humans killed wolves eagerly (both legally and illegally) because of their depredation of semidomesticated reindeer.[46] Some collared wolves dispersed long distances to the west. For instance, a tracked female wolf, originally from Sotkamo near the eastern border, traveled across the country and formed a territory with a mate on the west coast near Pyhäjoki in the autumn of 2002.[47] A male wolf called Noppe dispersed from Nurmes in Northern Karelia near the eastern border to the west coast, to Lapväärtti near the Baltic Sea, during the years 2003 and 2004. He traveled approximately 40 km a day. Noppe met a female wolf in that area and stayed there.[48] Most of the wolves living in Finland during my research period lived in eastern Finland. Researchers and officials—and some people living in the east—wished that more wolves would roam to the western and southern parts of the country.[49] With the help of tracking, researchers tried to find out how wolves managed to live in areas that were more densely inhabited by humans than eastern and northern Finland.[50] Researchers also studied wolf predation on wild forest reindeer in the Kainuu region. During radio- and snow-tracking in 1998–2000, a total of 467 fresh wolf scats were collected and analyzed to discover wolves' diet.[51] Wolves' avoidance of roads and human settlements was also studied with the help of collars.[52] These studies were possible due to the participation of wolves and the performative nature of technology.

INTRA-ACTION AND UNEXPECTED RESULTS

Barad uses the term "intra-action" to stress that agencies are entangled and not separate or distinct; they do not enter interactions as ready-made agencies but emerge in intra-actions and are mutually constituted.[53] According to Barad, in "technoscientific practices, the 'knower' does not stand in a relation of absolute externality to the natural world being investigated."[54] In wolf research collared wolves, researchers, and technological equipment were not independent objects, but, in Barad's way of thinking, they constituted a phenomenon in which knowledge on wolves was formed.

As Barad reminds us, technology can cause difficulties when it fails to work: "getting the instrumentation to work in a particular way for a particular purpose" is part of the difficulties of doing science.[55] Sometimes the technology used in wolf research refused to play along. Even though GPS collars were easier to use than radio collars, they needed to receive signals from at least three satellites to work. A wolf wearing any kind of collar needed to be in open terrain. Obstacles such as hilly terrain, moist vegetation, and buildings could block signal transmission. Ronkainen said in an interview in 1998 that if the collared animal was in a flat bog behind a tree-covered hill, the receiver would not get the signal. The reception could also be affected if the receiver was close to a person, vehicle, or the ground.[56] Unsuccessful locations delayed data sending, sometimes causing contact with some wolves to be lost. According to *Helsingin Sanomat*, in July 1998, when the tagged wolves had carried the collars for only a few months, the collar of the female wolf Jonna did not, for some reason, send any radio signals, and another female, Maija, was not located.[57]

The batteries of the collars lasted from under a year to two years. However, transmitters could fail before the end of their predicted operating life.[58] *Helsingin Sanomat* wrote in January 1999 that none of the five collars in use at that time worked with certainty.[59] The GPS receivers used in Finland from 2002 on required considerable power, and their operational lifetime could be shorter.[60] After the batteries died out, the wolves needed to be caught again and given new collars with fully charged batteries.[61] Researchers tried to learn from these technical difficulties: for instance, the wolf Noppe wore a "next-generation" collar, which had separate GPS, GSM, and VHF units so that the collar would continue to send data even if one of its units stopped working.[62] Dying batteries and failing transmitters tell about the vital role technology had in producing knowledge on Finnish wolves.

Beside technical problems, the animals equipped with collars might end up breaking them. Other wolves could chew the collars their pack members wore and break the cords.[63] A lone female wolf, Taiga, managed to take the collar off her neck in summer 1998.[64] This incident shows how wolves were intertwined agents in the research process alongside human researchers and technology. It also suggests that the wolves did not always carry the collars willingly. Because it was impossible to ask for the wolves' consent or explain the reasons for carrying a collar, they were all unaware that they were participants in research. If Taiga managed to remove her collar on purpose, because of discomfort, for instance, one could state that this act highlights her agency and capability to affect the research.

In the early days the collars were heavy: in 1969, researchers tested satellite-tracking in the United States and used a collar on an elk that weighed 25 lbs. The elk died only six days later.[65] Even though collars were smaller at the turn of the century, it is important to ponder how the wolves felt about wearing them. Also, the collaring procedure causes stress.[66] According to researchers, wolves are easy to anesthetize, and complications from overdosing are rare. Still, the anesthetizing causes a risk of hyperthermia, hypoventilation and other complications.[67]

Wildlife scientists have long considered the ethics of tracking. Finnish researchers needed to get permits in order to capture wolves, and they insisted that no harm or injury was inflicted on the wolves during or after the collaring.[68] Even though definitive tests of transmitter effects were rare, the researchers stated that the neck collars rarely affected survivorship, reproduction, behavior, or condition in medium- to large-bodied terrestrial mammals.[69] Altered behavior was not wanted because the observed animals were supposed to behave in a "normal" way. Wolf researchers did not want to enter into a responsive relationship with wolves, such as in the studies done with the primates that Haraway writes about.[70] According to Fuller and Fuller, a study fails or produces biased results if radio-collaring causes aberrant behavior or physiological stress, increases mortality, or reduces reproduction in the studied animals.[71]

Beside moral ones, radio-tracking raises questions about wild animals' individuality. Individual animals are often treated as representative of their species.[72] Nevertheless, the wolves who carried collars revealed something about wolf individuality. *Helsingin Sanomat* wrote in 1999 that the researchers had already found out "irregularities" when studying the wolves' diets, for instance, that different packs had different diets. Also some packs ignored domestic animals such as sheep and dogs even when they were present in their territory, while others specialized in preying on them.[73] The studied wolves often hunted elk alone and only rarely as a pack.[74] In addition wolves differed significantly from one another in their use of buffer zones around human construction. Despite the general belief that wolves are a safety threat to people and domestic animals, collared wolves were found to avoid human constructions such as buildings and roads in eastern Finland.[75] In Finland it had been believed that it is mostly male wolves who disperse over longer distances than females. However, the tagged wolves revealed a different story: distance did not differ significantly between sexes.[76] Thus the studied wolves could give unexpected information.[77]

Birke, Bryld, and Lykke state that laboratory rats are agents of their own history and active participants in the creation of their own meaning.[78] This argument could be challenged because humans can almost totally control the rats. It is easier to see wild collared wolves as agents of their own history. Technology even highlights this; Pschera writes that it "turns animals, which we are used to thinking of as objects, into subjects with their own backstory and fate."[79] According to Pschera, GPS-collar-wearing wolves "are no longer anonymous beasts that appear out of nowhere, attack, then vanish in the mist. Instead they have names and personal histories. They are no longer simply representatives of a species, but rather, real individuals. We can research their biographies and personal preferences as easily as their personalities and social behaviour. This ultimately transforms wolves into likable contemporaries—a remarkable career for the creatures, which have for centuries been pegged as humans' archenemies."[80]

By living their lives while wearing radio or GPS collars, the wolves revealed information about their behavior, preferences, and personalities as well as events from their life histories. For instance, researchers found out when new packs were formed in the research area, how many pups were born, and the number of juvenile wolves dispersing and not dispersing and where they ended up. Knowledge concerning reproduction was linked to the wolves' movement: female wolves were known to have given birth when they constantly stayed at their den site for three weeks.[81] This kind of information had previously been extremely hard to discover.

As historian Tina Loo reminds us, "we don't think of wild animals like grizzly bears as having a history, but they do, and not just collectively and evolutionarily, but as individuals."[82] Loo and field biologist and artist Colleen Campbell write about this in the context of collared grizzly bears whose lineage and family trees, in addition to their movement and actions, became visible to humans through tracking.[83] The research methods that bound technology, humans, and wolves closer together physically made it possible for the researchers to see clearly that each wolf was an individual. This was emphasized by naming the collared wolves: Taiga, Maija, Milla, Saturnus, Petrus, Ukri, Ugri, Igor, Noppe, and so on.[84] Collared animals were also named in the United States.[85] Naming is linked to ethics: it is harder to kill a named animal because a proper name humanizes them.[86] Some people living in wolf areas did not like the fact that collared wolves were given human names; they felt that the wolves got more attention than the people living in remote districts.[87]

People like hunters who wanted to know the whereabouts of collared wolf packs, were able to get this information from a telephone information service, Susipuhelin, from 2003 on.[88] The general public and people living in wolf areas could also read specific information about the wolves from newspapers and nonfiction books. The wolves' everyday lives—and even how many deer the packs killed—were described enthusiastically in a nonfiction book written by Erkki Pulliainen. He described how "Sotkamo's wolf couple Taiga and Petrus are specialized in Finnish forest reindeer. Mrs. Taiga hunted, by herself, six forest reindeer between May 30 and April 30, 1998. During spring 1999, the couple took five forest reindeer and one elk."[89] In 2001 an opinion piece stated in *Helsingin Sanomat* that after the wolves are collared, they are no longer just wolves but known individuals.[90]

One article in *Helsingin Sanomat* criticized the researchers' closeness with the studied wolves. The article described in a satirical way how the wolf expert Seppo Ronkainen showed photographs of the local radio-tagged wolves in a public event in Kuhmo in April 2000: "The slides showed Ukri sitting in a cage waiting to be collared, Maija's clear gaze, and Jonna with the cubs. It all looked so sweet, as if we were looking at the research expert's family album. One almost expected to see a picture of Maija taking a cruise to Tallinn or going to Tenerife."[91] This critique is an example of resistance to anthropomorphism, or what primatologist and ethologist Frans de Waal has called "anthropodenial," a process where we humans deny the fact that we too are animals and therefore share many characteristics with other animals.[92] Even though some people felt that the researchers and research assistants anthropomorphized the wolves, wild animals have backstories and personalities, and this is what their collars share with humans.[93]

Tagged wolves also revealed information about their deaths. According to *Helsingin Sanomat*, thirty-nine radio-tagged wolves had died from autumn 1998 to March 2004: sixteen were killed legally, some illegally, and the rest disappeared.[94] Researchers found out that all adolescent wolves who traveled to the reindeer herding area were shot before they reproduced.[95] Many of those who stayed away from the reindeer herding area were also killed. For instance, a tagged female wolf, Milla, and her mate were illegally killed in Pyhäjoki, North Ostrobothnia, in 2003.[96] Not all of the collared wolves were killed by humans: a female elk trying to protect her calf killed a four-year-old collared female in Sonkajärvi in 2004.[97] Information on wolf mortality in Finland at the turn of the twenty-first century was needed to make wolf conservation successful. As Fuller and Fuller state, it is important to know about death: "Unbiased estimates of survival and mortality of individuals leads to better understanding of population status and limiting factors."[98] Still, the deaths of collared wolves were a setback to Finnish wolf research.[99]

Wolf tracking revealed information about wolf individuals to people sharing their living space with these animals. During my research period, many people living in wolf areas in Finland were pleased about the radio-tracking and wished that more wolves would be collared. Still, not all local people liked the fact that wolves were collared. Jukka Bisi and Sami Kurki studied provincial and national expectations and objectives for the management of the Finnish wolf population and collected attitudes and views from almost two thousand Finns in 2004. According to them, some people believed collaring changed wolves' behavior and felt that the damages caused by collared wolves were somehow the research institute's responsibility. To some, tagged wolves were no longer wild and free but somehow controlled by humans.[100]

Knowing more about wolves did not necessarily make people feel more positively about them.[101] Some tagged wolves were purposefully hunted because they wore collars and participated in wolf research. *Helsingin Sanomat* wrote that according to the police, a collared pack leader called Ugri was killed deliberately. Two men from Suomussalmi, in Kainuu, were arrested. The newspaper wrote about the unreal thriller-like events leading to the wolf's death. Ugri had entered the reindeer herding area in Suomussalmi with his pack members. Wolves are not strictly protected nor are they wanted in the reindeer herding area. Due to the collar tagged to Ugri, researchers were aware of his whereabouts. With the help of a snowmobile, they got Ugri to run out of the reindeer herding area to Kuhmo. But Ugri did not run to safety. According to the newspaper, two men waited in a van for him a few kilometers from the gate. Ugri was run over unconscious, dragged into the van, and after few hours of driving in the Kainuu area, finally driven into a yard in Suomussalmi and killed with an electronic stunner used for killing furbearers. Ugri was then skinned and cut up. The fur and the body parts were hidden, but the police retrieved them and the Finnish Game Research Institute clarified the cause of death.[102]

Targeting collared wolves was an unexpected result of the research. A potential motive for this criminal act was suggested in *Helsingin Sanomat*: an interviewed reindeer owner stated that Kuhmo was the wrong place to conduct wolf research with radiotelemetry as Kuhmo was a neighboring municipality to Suomussalmi, where the reindeer herding area began. According to some locals, radio-tracking stressed the wolves. It caused them to roam to the north into the reindeer herding area and cause harm there.[103] Some Finns mistrusted wolf research. For instance, some hunters were skeptical about the population estimates and argued that there were in reality more wolves living in the country. Radiotelemetry did not dispel all of the existing mistrust, and it created new frictions between researchers and different interest groups.[104]

Still, some wolves escaped death because they wore a collar: A police officer almost shot Noppe, who with his mate had caused fear in Lapväärti after attacking a cow and some sheep. The officer had the right to shoot a wolf if people's safety was threatened, but after seeing the collar around Noppe's neck, he did not fire at Noppe but at his mate, whom he missed.[105] The collar saved Noppe's life in this incident.

PRODUCING KNOWLEDGE TOGETHER

In this chapter, I have shown how knowledge of wolves was a cocreation of technology, wolves, and humans. Research conducted with the help of radio and GPS collars tells a story about natureculture and multispecies participation. It tells us about physical and spatial intra-action, the blurring of objects and subjects involving humans and wolves, radio receivers, antennas and satellites, snowmobiles, syringes and drugs, computers, maps, and articles. It also tells us about wolf conservation and hunting, and the human urge to control wolves and their movements.

Collars can tell us much about the wolves' own history, the lives and deaths of the sixty wolves collared in Finland during 1998–2004 and their pack members. Pschera links tracking also to rights. After seeing and witnessing wild animal individuality, humans are required to grant them identity and see that "these are not species in need of protection" but individual animal subjects in need of personal rights. To Pschera the data stream constitutes the wild animals' right to exist.[106] In this chapter I have tried to show that wolves can be seen as agents participating in wolf research, cocreators of knowledge. Using new materialist notions of agency as inspiration for my analysis has enabled me to highlight that wolves are not passive objects or mere victims of human curiosity but active participants alongside the technology used in research. Still, we need to ask what was in it for the wolves? Is it fundamentally good or bad for people to know where animals are or how they live their lives? Pschera reminds us that if the positioning data of endangered animals falls into the hands of poachers, it can be used against them.[107]

As the bleak stories of illegally killed radio-collared wolves like Ugri tell, participating in wolf research could be dangerous for some wolves. After the wolves became known as individuals, some of them could be targeted. Nevertheless, many radio-collared wolves were admired and by carrying collars, they were able to help widen our knowledge of the species. As Coleman points out, humans know more about wolves today than at any other time.[108] Radiotelemetry and other methods of wolf research have helped to dismantle superstitions and update cultural notions, for instance, about wolves' interaction with prey, willingness to move about near human settlements, and the idea that all wolves are dangerous to dogs and sheep.[109] As Stephanie Rutherford has shown in the context of Canadian wolf history and Karen Jones in the United States, the cultural notions of wolves have gone through major changes during the last century.[110] Nevertheless, peoples' views on wolves are still conflicted.

Collared wolves have also helped people to learn ways to coexist with the species. From 2003 on, hunters and other people could inquire about the whereabouts of collared wolf packs using a telephone information service. These data were used, for instance, to prevent wolves from attacking hunting dogs because hunters knew which areas to avoid.[111] Therefore knowledge of the wolves' whereabouts helped people (and dogs) avoid them and prevent conflicts. To answer the question of what wolves got from the research, one could state that in a best-case scenario, the wolves that participated in field research helped people understand the species better, prevented conflicts, and enabled coexistence.

CHAPTER 7

TRANSFORMING THE HUMAN-EAGLE RELATIONSHIP THROUGH CONSERVATION TECHNOLOGIES

TUOMAS RÄSÄNEN

The Anthropocene manifests not just in the deteriorating environment and the destruction of the natural world but also with a multifaceted human attempt to control and limit the loss of species and biodiversity. While the populations of most wild fauna are in freefall, some previously endangered species have been able to bounce back and are indeed thriving. This has happened in particular in the affluent West, where attitudes to certain animal species have gone through a radical change in recent decades, enabling them to adapt to human-modified environments. A case in point is the white-tailed eagle in the northern Baltic Sea area, as the eagle population has soared from a mere few dozen breeding couples in the 1970s to more than one thousand pairs in the early 2020s.

The white-tailed eagle is the largest raptor in northern Europe. It used to breed in all coastal areas of the Baltic Sea and on the shores of several large inland lakes. It usually builds its nests in the tops of sturdy conifers, and the large twig nests can be hundreds of years old. It lays one to three eggs, normally in early spring. The species has a seasonal diet, of which its most important prey are fish. In the spring it also preys on different bird species, especially common varieties of waterfowl. In winter, when the northern Baltic Sea is mostly frozen and the majority of birds have migrated, the remaining eagles have to make do with scavenging carcasses (or nowadays with side catches left by fishers). The eagles use different tactics when trying to catch prey. Sometimes they fly above the archipelago without any obvious intention to hunt. In so doing it may lead a weak prey bird to become separated from a fleeing flock. The eagle then focuses on hunting this individual (thus acting as a selective force that improves the health of bird populations). Sometimes the eagles try to surprise prey by flying just above the tree line or by waiting patiently and inconspicuously in a tree for hours.

The revival of wild animal populations has often been a by-product of societal and attitudinal changes, which have, for example, reduced hunting pressures or made environments suitable for certain species. In some cases it has required years of painful and innovative work by conservationists, without which this positive turn would not have been possible. In northern Europe no other animal has been the subject of such intensive protection, in terms of both time and scale, as the white-tailed eagle. In this chapter I will examine the changing relationship between humans and the white-tailed eagle in Finland and Sweden. The first part of the chapter will review the decline of the eagle population prior to the 1970s, while the main focus will be on eagle conservation since the late 1960s.

There has been only one scholarly study on human relations with the white-tailed eagle, which examined the role of the birds in building awareness about toxic pollution in the Baltic marine environment in the 1960s and 1970s.[1] The story of the eagle has been told many times over the years by conservationists, as well as in numerous newspaper articles in Finland and Sweden.[2] These reminiscences have not only recapitulated the gradual recovery of eagles but also documented the work of conservationists. What is missing from these reports, however, is an analysis of the ideological and historical environmental contexts of the human-eagle relationship. I aim to fill this gap, with a special emphasis on technology as an agent of change in the human-eagle relationship.

The main source materials utilized for this study consist of documents (memos, letters, minutes, and public writings) of the World Wildlife Fund (WWF) Finland Working Group on the White-Tailed Eagle, which have not been previously cited in scholarly works. A tiny fraction of these documents has been published in newspapers, popular scientific magazines, and the journals of conservation societies, yet the vast majority are not even archived and are still in possession of the members of the working group. I have been able to study the documents gathered by Professor Henrik Wallgren, who acted as the chairman of the Finnish working group. This personal collection covers the period from 1972 (when the group was established) until 1995. Unfortunately, the documents related to the Swedish project (Projekt Havsörn) are at present not even organized, let alone available to researchers. Therefore I have been forced to rely on published material and discussions with Björn Helander, the longtime head of the Swedish project.

CARE PROTECTION

The history of the human-wildlife relationship has repeatedly been portrayed as a battle of ideas about and the politics of nature, followed by the practices that subsequently emerged. In these narratives conservation has either appeared from scientific ideas of aesthetics, culture, and the scientific or economic bene-fits of healthy populations, or from moral obligations humans have felt to their nonhuman cousins.[3] Much less attention has been given to the technology that enables us to relate to and acquire knowledge on nonhuman beings, from the tiniest pathogens to the creatures that inhabit the depths of the ocean. Conser-vation always requires technology of some sort.

The emphasis on ideas and politics is understandable when considering tra-ditional conservation methods, in which a certain species is just protected from hunting or habitats are preserved from land use that is harmful to them. That said, technology can of course be a powerful asset in defining the specifics of traditional conservation as well as in preventing violations. However, Swedish and Finnish conservationists quickly realized that such a traditional approach to conservation was utterly insufficient. This led them to innovate new conser-vation techniques to protect white-tailed eagles, such as winter feeding and the building of artificial nests.

I argue that Swedish and Finnish conservationists developed a wholly new type of conservation through such techniques, which I call *care protection*. I un-derstand "care" in the sense of a proactive interest in the welfare of other beings, which manifests itself in the form of working for the other. Care is a reciprocal process that not only brings benefits and comfort to the one who is cared for, but also satisfaction and pride to the caregiver for being able to support the other. Care is often understood as something that happens between two human beings who are physically close to each other, a prime example being the rela-tionship between nurse and patient. However, care can also take place between a human and an animal. Similar to human-human relationships, an animal, in its relationship with a human, is not necessarily a passive object of that care. As will be seen later, although humans undertook care work, eagles were active actors in constructing an interspecific care relationship. Moreover, care can also occur across space, such as between two actors, one of whom lives in a city and the other in a remote archipelago island. More important than distance is the emotional affection and work of the caretaker toward the other.[4] The term "care protection" derived from the fact that the care of individual eagles aimed to protect the species from becoming extinct in the Baltic Sea area.

It is not surprising that care protection was first implemented with regard to birds. After all, there is a long global history of protecting birds. Many of the first animal conservation movements were established to protect birds, and humans have hung nest boxes for birds for centuries.[5]

Care protection of the white-tailed eagle officially began in Sweden in 1971 and in Finland in 1972, which to the best of my knowledge also marked the first of its kind in the world. Since then, although often without any reference to or influence from the protection efforts for the white-tailed eagle, care protection has been adapted to numerous global conservation projects, from leatherback sea turtles and Saimaa ringed seals to freshwater pearl mussels and many other critically endangered species. However, I do not know of any other similar activities pertaining to the protection of endangered animals that predate the protection of the white-tailed eagle. It is true that humans have long been managing populations of game animals by, for example, feeding them.[6] However, there is a clear difference between protecting an endangered animal merely because it is endangered and feeding it in order to be able to slaughter it.

By early 1970s the protection of endangered species was becoming an important issue and gaining a lot of momentum. WWF, for example, supported more than one hundred different conservation projects worldwide since the inception of the protection of the white-tailed eagle. These included establishing nature reserves to hiring guards for protected areas, from funding population assessments to organizing conferences. None of these activities, which I have termed "care conservation," appear to have been carried out.[7] The focus in traditional conservation has been on population levels at the expense of emphasis on individual animals.[8] Care protection has placed the nurturing of individual animals at the center of its action. As such, it somewhat resembles the rescue and rewilding technique, in which endangered animals are raised in captivity and later reintroduced into the wild. Some rescue and rewilding operations predate the conservation of white-tailed eagle.[9] The novelty of care protection and the distinction between care protection and rescue and rewilding operations arises, on the one hand, from the many ways in which care protection takes place, but especially from the fact that care protection aims to protect wild animals in situ, that is, in their own habitat.

The technology used by conservationists in the 1970s and 1980s may have been rudimentary in comparison to present-day satellite tracking systems and online devices. Nevertheless I would argue that it provided them with the necessary means to enact care protection. Technology was the mediator for conservationists in a Latourian sense: it enabled them to overcome the space between city-dwelling conservationists and island-dwelling eagles and helped to further knowledge about the eagles. This led to the formation of a bond between humans and these animals.[10] What is more, as conservationists actively sought to assist the survival of eagles, they also had to take into account the agency of the eagles themselves and their life histories. Consequently, and with the support of the technology used in protection practices, conservationists began to conceive of eagles as individuals, every one of which was important to the entire population.

RIFLES, TOXINS, AND THE DECLINE OF THE EAGLE POPULATION

The white-tailed eagle is an apex predator in the archipelagoes and coastal areas of the Baltic Sea. Their primary prey of fish and waterfowl are also valued in human diets, and hence eagles have been treated as competitors for centuries.

In addition, eagles sometimes kill domestic animals, and in folktales they were described as eating little children, which aggravated humans' fear and hatred of the birds. In premodern times the sparsely inhabited coastal areas ensured that these attitudes did not significantly impact the eagle population. As the human population rapidly grew in the nineteenth century, at a time when Baltic-area governments tightened their grip on natural resources within their territories, predators were perceived as impeding development and progress. Thus they came to be viewed as enemies of civilization. From the 1870s the eradication of the white-tailed eagle in coastal areas became the most important game management goal, which was encouraged by offering bounties to hunters.[11]

Administrations in the Baltic area provided both a justification and an incentive to get rid of eagles. Firearms provided the means. Rifles manufactured in factories began to replace homemade rifles by the late nineteenth century. These guns were not only of better quality, but were also relatively cheap and well within the means of recreational hunters as well as the local populace.[12] It was, of course, people, not guns, that killed eagles, but people without guns killed many fewer animals. In only a few decades a healthy eagle population of several hundred pairs diminished to less than a hundred pairs in the northern Baltic Sea area by the 1920s. Moreover, the breeding population was scattered to various separate enclaves.

The white-tailed eagle was officially protected in Sweden and the Finnish Åland Islands in 1924 and in the rest of Finland two years later. This did not do a great deal to benefit the bird population, as the governments had neither the will nor the resources to fully enforce the law. Animosity to the eagle continued to be a factor. Indeed, eagles were frequently shot on sight, their nests destroyed, or their eggs taken for wealthy collectors in Europe. The persecution suffered by these eagles ensured that they learned to avoid humans as much as possible. The few surviving eagles nested in the outer archipelagoes, where human encounters were rare.[13]

The killing of eagles continued until the latter part of the twentieth century. However, it was toxic chemicals that finally almost drove the white-tailed eagle population in the Baltic Sea area into extinction. Swedish ornithologists had already noticed in the late 1950s that the number of fledgling eagles was declining, but at the time no one realized that chemicals, such as DDT, PCB-compounds, and mercury that were used in agriculture, forestry, and industry were able to travel to the sea and up food chains into the tissues of eagles. By the turn of the 1970s, when the impact of chemicals spreading into the environment and the mechanisms through which they entered into animals became understood, the eagle population had plunged from maybe a little over a hundred pairs in the 1950s to a few dozen in both Finland and Sweden.

Toxic chemicals even made adult eagles vulnerable to death during severe winters.[14] Even more worrisome was the fact that hardly any nestlings hatched, and the population was aging. Between 1970 and 1975, for example, only three nestlings in the Stockholm archipelago survived. In Finland the situation was almost as bad. During the worst year, 1973, only five nestlings hatched. Chemicals made many eagles virtually sterile. Despite this, some eagles still succeeded in producing eggs, but most of the time they did not lead to healthy fledglings because hormone imbalances caused by toxic chemicals made the eggshells too thin and porous; they often broke prematurely. Ornithologists also found eggs that looked normal but were as light as "ping pong balls," as stated by Björn Helander, because the embryo had dried up and disintegrated. It was estimated that a success rate of 50 percent was needed to keep the eagle population stable, yet at the time the reproduction rate was barely 20 percent.[15] It seemed that the eagle population was inevitably facing a downward spiral, as old eagles would gradually die off, and there were not enough thriving fledglings to replace them.

FROM GAS CHROMATOGRAPHY TO A NETWORK OF PROTECTION

The reason for the low reproduction rate among the eagles would have been inexplicable had there not been sophisticated apparatuses available to detect minute amounts of chemical substances within the avian creatures. A method of coupling gas chromatography to mass spectrometry, which was invented in the 1950s, was cultivated by scientists at the Swedish Karolinska Institut and proved to be extremely useful in enabling toxicologists to begin to disentangle the toxic bombardment of the Swedish environment.

While studying DDT pollution in the Baltic marine environment in 1965, Sören Jensen, a Danish toxicologist working at the University of Stockholm, found unknown peaks in his gas chromatography analysis. The first of these puzzling peaks was drawn from samples from a white-tailed eagle that had been brought to the Swedish Museum of Natural History in 1944, that is, before DDT was even used in Sweden. In later studies Jensen determined that it was PCB, not DDT, that had caused the peak. The combination of advanced technology and the eagle specimen that had been found dead more than twenty years before he began his research helped Jensen make the initial discovery of PCB compounds in the natural environment, which he publicized as a short report in *New Scientist* in 1966.[16]

It was not a coincidence that Jensen's discovery occurred in the mid-1960s. Following a public outcry about toxic chemical pollution, the Swedish government allocated plenty of resources for toxicological research. The results of these studies shocked the authorities when they were published a few years later. Organisms from the Baltic Sea were discovered to have five to ten times more DDT and PCB compounds than any other marine area in the world. Furthermore, the levels of lethal methyl mercury were almost as high as in Minamata, Japan, where hundreds of people had died from mercury poisoning. As these substances concentrated up the food chain, it became clear why eagles, which occupied its upper echelon, had staggering amounts of toxins in their tissues.[17] These astonishingly high levels galvanized environmentalists in Sweden and Finland and ensured that the Baltic Sea area became the focal point of environmental discourse in both countries.

However, the new environmental advocacy that emerged in the 1960s and 1970s was not concerned merely with securing an unpolluted environment for humans. It also extended of the scope of classical preservation, in which the main focus had been on national landscapes and cultural heritage. Thus at the same time that groups advocating the well-being of the human environment grew, it was also noticeable that there was heightened interest in undisturbed natural environments, endangered animals (many of which had been hunted because they were predators), and, in more general terms, healthy biotic communities. For example, the WWF, established in 1961, gained a lot of publicity and new members by the end of the decade, which enabled it to expand its conservation activities.[18] Thus it was the coalescence of these environmental factors, rather than a single factor, that led to growing interest in the white-tailed eagle among conservationists in Finland and Sweden.

Prior to the late 1960s only individual scientists had undertaken surveys of the eagles and campaigned on their behalf. Now a younger generation of nature lovers and amateur ornithologists from coastal areas stepped in to take the lead in the fight to save the eagles. According to their own testimony, many had been inspired as young boys by Bengt Berg's stories from the 1920s about the white-tailed eagle in Swedish nature.[19] In Sweden Projekt Havsörn began in 1971 under the wings of the Swedish Society for Nature Conservation (Svenska Naturskyddsföreningen) and was led by the young zoologist Björn Helander. In Finland, the Working Group on the White-Tailed Eagle, founded in 1972, operated under the auspices of the newly established Finnish branch of the WWF. The Finnish working group was chaired by Henrik Wallgren, a professor of zoology at the University of Helsinki. However, the conservation praxis was mainly carried out by other, younger members of the working group, consisting of some half-dozen ornithologists who would over the years become the leading experts in the study of white-tailed eagles in Finland and beyond. In both projects the main actors in the working groups, who themselves did the conservation on a voluntary basis alongside their work or studies at research institutes, were accompanied by dozens of local volunteers who helped to implement the goals of the project. From the very beginning, the Swedish and Finnish projects collaborated closely.

The idea of protecting eagles may seem far from extraordinary, even old-fashioned, when compared to the current trends in conservation biology. After all, it relied on a traditional species-by-species approach, which the leading scientific voices of the 1970s deemed insufficient, instead advocating for the conservation of a diversity of species, genetic traits and ecosystems.[20] The novelty of the approach to white-tailed eagle conservation was in its practices that incorporated a complex set of techniques.

Traditional land-use restrictions were still part of the conservationist's playbook. For example, they repeatedly tried to purchase land in order to establish designated eagle conservation areas. Landowners were often reluctant to sell, so conservationists pressured them and forest management officials to refrain from pursuing destructive land-use practices in the vicinity of eagles' nests.[21] The white-tailed eagle was already protected by law, however, and therefore merely dedicating nesting areas for eagle conservation would not have made much sense, since the birds roamed far and wide in search for food. Furthermore, no matter how large a preserved area was, eagles would still consume toxic fish and waterfowl. Consequently conservationists had to invent new techniques in order to entertain the slightest chance of saving the eagles. The greatest area of concern in this regard centered on winter feeding. The purpose of promoting winter feeding was to provide young eagles with uncontaminated food during their first winter, when they were most vulnerable to starvation. It was also hoped that feeding eagles in the Finnish and Swedish archipelagoes would encourage them not to migrate south, where the chance of being shot was even greater and where they would only find toxic food. It was hoped that the winter-feeding initiative would help to nurture them safely into adulthood. Moreover, by the time the toxic contents in the environment had diminished, in the wake of the banning of the most notorious toxic chemicals in the early 1970s, the birds would be able to produce healthy offspring. The winter-feeding initiative began modestly in the late 1960s, but intensified with the projects' official status in Sweden in the winter of 1971–1972 and in Finland a year later.

The winter-feeding initiative was a real tour de force by the conservationists and was carried out for more than twenty years. More than fifty winter-feeding sites quickly emerged along the Swedish coast, alongside twenty-five sites in Finland. Conservationists in both countries provided the eagles with up to 30 tons of pig carcasses (to which the Swedish even added vitamins) that they had received from slaughterhouses and piggeries.[22] Feeding wild animals is an old practice in game management, but it had never before been carried out in Finland or Sweden in order to protect an endangered species.

The history of animal conservation provides many examples of erroneous methods of feeding that have led, for example, to behavioral problems in animals as well as clashes between competing human interests. Species that have suffered from such misconceived feeding methods include California condors in the United States and brown bears in Finland.[23] The white-tailed eagle projects avoided such problems by establishing dozens of different feeding sites all around the Finnish and Swedish archipelagoes. Eagles at these sites never became too accustomed to humans or conditioned to wait and rely entirely on humans for their food.

CHAINSAWS, SNOWMOBILES, AND EAGLE AGENCY

The historian of animals Noah Cincinnati has convincingly argued that animal behavior played a pivotal role in the development of in situ protection in the 1920s. He shows how gorillas lost their will to live in captivity, which made scientists realize their rarity and the necessity of in-situ protection in order to save them from extinction.[24] In the development of care protection, white-tailed eagles were agents in a manner not too dissimilar to the gorillas that had been observed in the 1920s; their actions defined the way in which Finnish and Swedish conservationists imagined the options they had to save these rare birds. I pointed out how young eagles often decided to migrate south, thereby putting them in more danger. Conservationists aimed to influence this decision-making process by tempting the birds with surplus food and thus giving them an opportunity to stay in the icy north throughout the winter. For them, eagles clearly were agential beings, and that agency had to be taken into account when planning for their conservation.

Toxic chemicals were not the only problem that endangered the lives and successful nesting of white-tailed eagles. The end of World War II until the 1970s was a time of rapid social and economic development in the Nordic countries. The explosive human impact on the environment since the 1950s has been dubbed "the great acceleration" in the literature on environmental change.[25] In Finland and Sweden this is perhaps best seen in the insatiable exploitation of the bountiful forest reserves during this time. The 1960s and 1970s may be mostly remembered by the public in these countries, in environmental terms, for the immense areas in the Northern wilderness that were clear-cut. Industrial-scale logging did not spare the archipelagoes either. This process entailed the removal of wood using chainsaws until the 1970s and harvesters thereafter. In contrast to the labor-intensive work of the peasant forester and his horse in earlier times, chainsaws and harvesters enabled large areas to be cut down in a matter of a few days. It also led to the construction of a tight network of forest roads through which loggers entered the woods and timber was transported to wood-processing plants. These roads also lured other visitors, many of whom, thanks to rapid economic growth, had money to build summer cottages and purchase motorboats and snowmobiles. In Finnish coastal areas alone, for example, there were more than sixty thousand motorboats by 1970, a figure ten times higher than in 1950.[26]

Due to long-lasting human aggression white-tailed eagles have evolved to be fearful of even relatively minor human harassment, whether it is intentional or not. During the nesting season, in particular, they flee far away whenever humans try to approach their nests, leaving their eggs vulnerable to ravens and crows. Alongside the toxicity of the eagles' food, intensified human activity— whether it was logging, road building, or traffic on land, sea, and ice—became the single most important concern for conservationists. According to one expert, the harassment of eagles was deemed to be such a grave problem that "the dropping of nests or felling of nesting trees, which previously was an absolutely condemnable act, has nowadays become one of the best ways to protect the white-tailed eagle. Oh the times, oh the protection methods."[27] Although this is arguably an exaggeration, the statement testifies to the sense of frustration felt by conservationists when they repeatedly encountered a failed nesting site that had been caused by a local inhabitant who had collected wood for his or her home stove, or an ice fisher who had gotten too close to the nest with their snowmobile, or a curious nature-goer who had tried to catch a glimpse of a bird of prey.[28]

To circumvent the eagles' sensitivity to human presence, nesting areas, as well as winter-feeding sites, were kept secret in order to try to minimize visits by overly curious observers, as well as those who wanted to harm the eagles. Conservationists also tried to enlist sympathetic locals to guard the nesting areas, but this was soon abandoned as it proved ineffective.[29] When the eagle population numbered only a few dozen and reproduction rates were dramatically below normal, every successful nesting and every egg counted. Hence conservationists tried to secure the best possible nesting results by building artificial nests. The construction of artificial nests aimed again to affect eagle agency. Instead of eagles staying in their traditional sites, where nesting was likely to fail, it was hoped that the artificial nests would instead "guide eagles to more peaceful territories."[30]

The first artificial nests were built in Sweden in 1973, and a few years later Finnish conservationists followed the example of their Nordic neighbor.[31] To secure successful hatching, Swedish conservationists also experimented with incubation machines, after which hatchlings were placed in a nest in order to be cared for by foster parents.[32] This was apparently just an experiment as it never became a widely used method, whereas artificial nests became of great importance. In the Kvarken region on the Finnish west coast, for example, more than half of the area's twenty-five eagle pairs used artificial nests in 1994.[33] The eagles had indeed altered their behavior, according to the wishes of the conservationists. It is impossible to say whether choosing human-made nests, instead of the large nests they had previously used for decades, was related to more general changes in eagle agency that had been occurring since the 1990s. As hatred of eagles waned, the birds began to increasingly tolerate human presence and displayed a "wider amplitude in their choice of nesting places."[34] Artificial nests nevertheless were a prime example of care protection and of the substantial toil conservationists undertook without receiving any material gain. They were motivated only by the sense of pride and satisfaction in helping their beloved birds.

CONSERVATION TECHNOLOGIES AND THE INDIVIDUALIZATION OF EAGLES

Conservationists would have been well-off without having to spend day after day working to preserve eagles. However, ecological circumstances, such as toxic chemicals and ever-increasing human activity, left them with no other option if they wanted to save the birds that they had established a deep affection for. Care protection was therefore born out of necessity. Yet conservationists also needed tools in order to successfully execute this tactic. Consider, for example, the extraordinary difficulties associated with delivering thousands of tons of food to eagles that was necessary just a few decades earlier.

Not that it was necessarily easy in the 1970s. It is rarely mentioned in the documents, but the logistics of getting meat to the eagles in their habits were extremely complex. Moreover, the conservationists were working on a voluntary basis. Of course, roads were required. The same roads that were despised by the conservationists because of the disturbances they brought provided a valuable benefit that had not been afforded to their predecessors. Pig carcasses were often transported to the seashore in the backseat of a car.[35] The remainder of the trip—from the seashore to the archipelago islands—had to be carried out by skis and sledges, or by boats when the sea was ice-free. Luckily they soon began to receive donations, in the form of equipment or money, that could be used to purchase cars, trailers, snowmobiles, and boats as well as fuel. The Finnish Working Group on the White-Tailed Eagle, for example, received fuel from Neste, Finland's largest oil refinery company, and they also received large donations from the vehicle manufacturer Scania, among others. Finally, they needed refrigerated warehouses in order to preserve the food for the eagles.[36]

As noted in the working group's circular letter, "efficient protection is only possible if the living range and behavior of eagles are accurately known."[37] Aerial flights surveys offered a convenient and exciting means to gain such knowledge. After signing a contract with the Finnish air force and border guard agency, conservationists were allowed to participate in routine flights on small aircraft and helicopters, from which they were able to easily "discover new nests and new [eagle] pairs in the area . . . which sped up the protection."[38] Nonetheless surveillance flights only constituted a tiny fraction of the research. More commonly the only way to undertake observations was to spend hundreds of hours throughout the year in bird hides or by undertaking observations by climbing up trees in order to band young eagles, irrespective of the weather.[39]

The close and lengthy observations undertaken to protect eagles necessarily changed the relationship between humans and eagles. The most profound change in approach at this time may have been the manner in which eagles became individualized. On the one hand, as every specimen was of great value, sources sometimes reveal the emotional attachment of the conservationists to the eagles. Injured eagles, for example, were treated with great care in the hope of returning them to the wild. When an eagle was found with a gunshot wound, conservationists would launch a careful study that resembled a criminal investigation. Juhani Koivusaari, a working group member from the Kvarken area, for example, spent days, with the help of veterinary experts, deciphering the exact chain of events that had led to the death of an eagle. In press releases he made sure to refer to the dead eagle as a "six-year-old individual" and as a valuable member of a critically endangered species, indicating that this was both an extreme violation of the eagle's individual rights and a serious blow to Finnish nature.[40]

On the other hand, the conservationists gradually got to know the life histories of individual eagles. In 1976 Swedish and Finnish conservationists began to attach colored bands to eagle hatchlings.[41] Since it was impossible to recapture the birds as adults without injuring them, colored bands would have been of no use without the utilization of a technology that was able to read them from a distance. With telescopes, automatic game cameras, and later video and digital cameras, however, it was possible to study the annual and lifetime migration patterns of individual eagles. Scales placed in winter feeding sites measured how much meat the eagles were eating. Thus conservationists were able to connect the pieces of knowledge that technology helped to gather in order to gradually construct a clear pattern of the eagles' behavior and population dynamics. Although it was never possible to gather an all-inclusive picture, it nevertheless helped to view the birds as individuals with distinctive life histories.[42]

WORKING FOR THE EAGLE ON MANY FRONTS

The history of human-animal relations mostly consists of interspecific colonization and exploitation. Animals have been valued when they are useful, either by undertaking work or by providing nutrition. The white-tailed eagle does not conform to the above criteria. On the contrary, in terms of conservation it was humans who sacrificed a huge amount of time and energy for the benefit of the eagles. In this sense, care protection meant a radical volte-face in the history of human-animal relations.

Not all the work took place in the remote islands. In fact, conservationists labored on two different fronts. On the one hand, they did everything in their power to help the white-tailed eagle via active conservation methods. On the other, they tried to cultivate sympathy among the public toward eagles and the conservation of nature. For years it seemed that the whole project was merely a fool's errand. Despite all the hard work, most eagles failed to nest over the course of many years, and the eagle population remained frustratingly low. Little by little, however, both fronts began to receive increasingly positive signs. In the late 1980s conservationists were finally able to rejoice that "the population is no longer at risk of acute extinction."[43] This was due not only to the eagle project, as restrictions on the use of toxic chemicals from the late 1960s on finally started to have an impact on the marine environment, with a reduction in the levels of toxins. It is noteworthy, though, that it was precisely the findings of conservationists and environmental scientists on the accumulation of toxins in marine organisms that played a major role in raising awareness of these toxins.[44]

Meanwhile, donations to the projects began to grow in the 1980s and attitudes grew more favorable to the eagles. Changes in attitude "particularly concerned coastal and archipelagic inhabitants in the areas where the white-tailed eagle nested," as Henrik Wallgren, the chair of the working group, put it in 1993.[45] With the modernization of society, archipelagic nature and eagle habitat had transformed from a space for subsistence and living to a space for recreation, where urban people could find respite in wild nature. The majestic bird of prey soaring in the sky was a well-fitting symbol of this wildness and authenticity, not least because conservationists had already for years portrayed it as such. It was therefore convenient, and perhaps also profitable, for many companies presenting themselves as environmentally responsible actors to jump on board to support conservation.

Can the care protection of eagles be viewed as a success story, which benefited all the parties involved? As María Puig de la Bellacasa, among others, has reminded us, care can bring about both benefits and harm to those being cared for.[46] Speaking exclusively about humans and eagles, it is difficult to find any downsides stemming from the conservation project. Eagles, for example, have not become too accustomed to artificial feeding, which could have impaired their ability to survive in the wild, as has been a problem in some other feeding projects. Nor were any individuals harmed for the sake of the population, as has been the case elsewhere.[47] Humans, for their part, have almost unanimously welcomed the return of the white-tailed eagle. But in the shared ecosystem there is no such thing as "exclusivity." The care of eagles has not passed without unintentional consequences for the ecosystem. To cite just one example, as the number of eagles has multiplied, the population of common eiders, the eagles' typical prey, has significantly diminished. Today the common eider in the Baltic Sea is considered to be endangered according the International Union for the Conservation of Nature's Red List. The eagles are definitely not the only culprits, as eiders are also threatened by numerous anthropogenic factors, but they have played a part in the decline of the species. As is too often the case, attempts by humans to control nature, no matter how well intended, have created other problems within the ecosystem and among the species that inhabit it.

The conservation of eagles was an interactional process in which both humans and eagles participated in the making of a new interspecific relationship. This was characterized by the invention, or "becoming," of care protection. The concept of "becoming with" has been introduced by Donna Haraway and has been used in numerous works on human-animal studies to describe the changes that has been shaped in interspecific relationship.[48] It is often presented in introductions as a catchphrase without readers being told the results of that particular becoming. In the case of the Baltic Sea area, the affection and care of individual animals and, more specifically, even some new conservation technologies, such as winter feeding and the building of artificial nests, were concrete realizations of interspecific becomings.

In Latourian terms, socio-technical assemblages together with eagle agency made up care protection. Without the use of current technology, along with a sensitivity to eagle agency, conservationists would have never been able to get to know the habits of eagles, let alone come to recognize them as individuals, which were both prerequisites for the intensive care and the becoming it signified.[49] Thus while the demise of the white-tailed eagle population had been furthered by technological progress, technology also connected humans and eagles and played a vital role in ultimately protecting the species. The Baltic Sea, which was a space of conflict only a few decades ago, has "become" a shared space of coexistence and peace for humans and eagles. This brings a glimpse of hope at a time when animals are becoming extinct at an accelerating rate.

CHAPTER 8

CAN YOU SHIP A PENGUIN COD?

ELLEN ARNOLD

On November 7, 1939, Mr. J. W. Yates of Atlanta, Georgia, wrote to Admiral Richard E. Byrd, the famed polar explorer. He asked that as "a great lover of birds" he be allowed "to make an application for one of the twenty penguins which you have so generously offered to the public." He was sure that "there is a great demand for these Antarctic fowl," but still earnestly wrote, "If it were within my power to offer you any financial assistance in consideration for this bird I would not hesitate in doing so but due to the fact that finances are a limited thing with we Southerners, all that I can offer is a Southerner's undying love for nature and her children. *However, I am in a position to pay for the bird's transportation and a shipment c.o.d. by Railway express* to 204 Lawyers Building, Raleigh, N.C. will receive my immediate attention."[1] This is but one of a series of letters sent in November 1939 to the offices of Richard E. Byrd in response to an Associated Press (AP) bulletin that had been broadcast across the nation.

Byrd had recently exhibited penguins at the 1939 World's Fair in New York City. The exhibit had had mixed results: it had succeeded in exposing a national audience to still-rare penguins, but left Byrd financially stretched right when he needed to fund his upcoming Antarctic expedition. The exhibit had cost far more than anticipated, and in an effort to recoup some of the costs, Byrd decided to work with both the National Aquarium and his staff to sell the remaining twenty birds that had been on exhibit. However, the AP announced to the nation that Byrd was looking for "good homes" for the penguins. This led to a flood of letters and telegrams, which arrived in such astonishing numbers that the expedition wound up being delayed for several days as Byrd and his harried staff dealt with the remarkable and completely unanticipated volume of correspondence.

The events that led up to this bulletin, and the overwhelming response Byrd and his team received, provide a window into the ways that communication and transportation technology opened up spaces in interwar America for people across the country and from a wide range of socioeconomic backgrounds to learn about, experience, and encounter penguins, often for the very first time. There had been many failed efforts to bring penguins north of the equator, and even after the first penguins arrived in America, there were continued problems with keeping them healthy and getting them established in zoos across the nation. But as these obstacles were haltingly overcome, and as technical and zoological expertise grew, penguins could be more easily transported and cared for, meaning that their display in zoos, aquariums, and events like the World's Fair was much easier, though many penguins did not survive transit and subsequent captivity. By 1939 the plucky birds were increasingly a part of the American environmental imagination.

This is a story about how people were able to transport and transplant penguins and about how film and radio gave more Americans an understanding of the places that penguins normally inhabited. The display of penguins was part of the story of the modernizing nation-state and about efforts to control, map, and colonize Antarctica. The glorification of explorers, science, and American expansion intersected in Richard Byrd, an explorer, aviator, naval hero, and celebrity all rolled into one. Byrd established a polar base called Little America that, among other things, had a radio station and sent mail. This wrapped the Antarctic into the American telecommunications web and helped fuel interest (and funding) for polar science and exploration. So when Byrd found himself with twenty expensive polar birds on his hands, the intersection of his fame, growing awareness of penguins, and the power of radio to erase distance and flatten space led to the displacement of Byrd's birds in multiple ways. What follows is the story of how these penguins got to New York, how the American public encountered and learned about penguins in general and these penguins in particular, and why it was so easy for so many Americans to suppose that their places, too, could become penguin places.

"WHO WANTS 20 PERKY PENGUINS?"

The Admiral Richard E. Byrd papers at The Ohio State University has a box with a series of folders labeled simply "Penguin Requests."[2] These folders contain over 840 letters from across the United States. All were written or sent between November 5 and 11, 1939, and all appear to have been in response to the AP bulletin that went out across the telegram service and was almost immediately published by local and regional papers nationwide and reported on the radio.

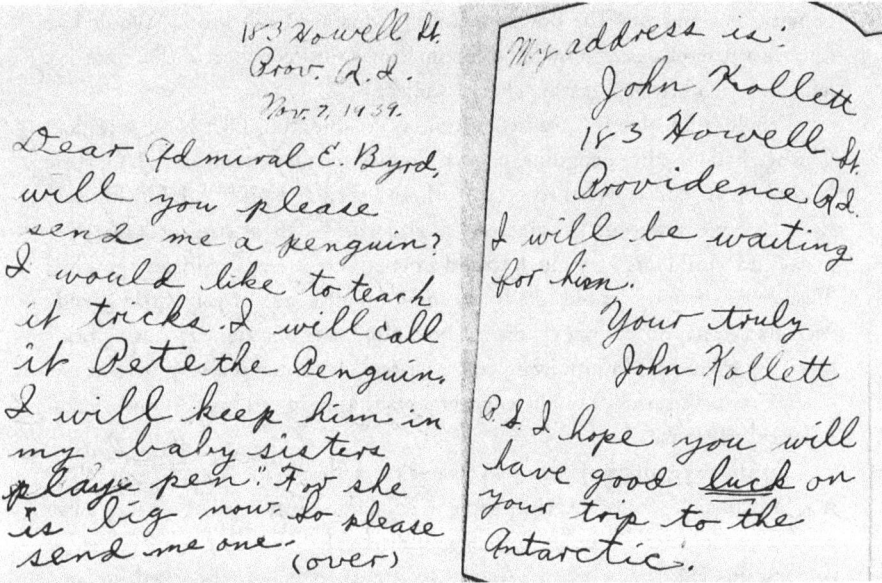

FIGURE 8.1. A sample from the collection of letters requesting penguins in the Byrd Papers. Source: Box 19, Papers of Admiral Richard E. Byrd, Byrd Polar and Climate Research Center Archival Program, The Ohio State University, Columbus.

A November 7 item from the *Washington News* read in part, "Wanted: good homes for 20 penguins. Will make fine pets, learn tricks, and follow you around. Apply Rear Admiral Richard E. Byrd."[3] Another asked, "Who wants 20 perky penguins?"[4]

Many of the letter writers explain how they had encountered the news. Some heard it on the radio, others read it in local papers, and some even included clippings from their local papers along with their requests. Most seem to have responded almost immediately. The very short time window in which all of this occurred was also facilitated by the modernization of postal services, the increased reliance on the telegram, and, most probably, the increased national attention to news as World War II dawned.

The deluge of penguin requests was also due to the fact that by 1939 penguins had waddled their way into the American psyche. Letter writers told Byrd about their collections of penguin memorabilia (including salt shakers, notepaper, and wooden toys), others drew him pictures, and some young children were able, not hyperbolically, to claim that they had loved penguins their entire life. By 1939 there were universities, fraternities, elementary schools, and girl scout troops who chose penguin mascots. The president of Whitman, Massachusetts's, Purple Penguin Skating Club asked for a penguin, and the members of University of Michigan's Zeta of Alpha Kappa Lambda wanted to adopt a penguin as a mascot.[5] The Bowdoin College Mustard Club wrote, "Would like to have a penguin as a companion for the Bowdoin Polar bear if there are any left please send transportation charges collect."[6]

People from North Carolina, Virginia, Washington, DC, New York, and Chicago had seen live penguins in zoos. Countless others, such as G. J. Guthrie Nicholson Jr. had traveled to New York and seen the polar exhibit. He told Byrd that he "saw your pengins [sic] at the worlds fair at present have a pet goose and would like very much to add a penguin you very kindly autographed 'Discovery' for me I would like to do something for one of your little friends very glad to pay shipping expenses."[7] Nicholson was not alone in noting prior meetings with the admiral. Many letter writers recounted meeting Byrd at one of the extensive series of public engagements that followed his first and second polar expeditions.[8]

Yet the exposure to penguins was not yet widespread. A man from Algoma, Wisconsin, noted that "the people in my community have never seen a live one." Two Colgate University students admitted that "Frankly, our knowledge of penguins is rather limited." And at least two letter writers asked for "one of the pelicans I understand you are giving away," and another wanted "one of your Pelegans."[9] Because of ease of communications, the overall sense of the reliability of postal services, and the fact that despite the obvious fact that few people had more than what one writer termed a "layman's knowledge" of the birds meant that they took for granted the fact that penguins could be easily delivered to them across the United States.

Meredith Masters from Springfield, Illinois, who was "eleven years old, going on twelve" was also concerned with the care of the penguins, asking if she could keep a penguin in her rabbit cage (which she helpfully drew), and wondering "How much will it eat? Does it have to have a pond? Do you think it will live in this climate?" She told the Admiral, "I will pay to have it sent and for it's [sic] food and care. Will you let me know the price?"[10] Young Ann Lewis of Granville, Ohio, was similarly inquisitive, asking, "How much do they cost? What do they eat? How do you take care of them? Can they ride in cars? How old are they and how much longer will they live?"[11] Intriguingly, it was often the youngest correspondents who admitted their lack of knowledge. Most adults simply assumed that the penguins could be shipped, and that the transportation would be affordable.

These assumptions are most visible in the telegrams, by their nature more terse. Irvin D. Slusser of Waukegan, IL, wired bluntly, "Please send one or all penguins." Paul Lackie of Lawrence, Kansas, wired: "Have good home for young male penguin if you can ship prepaid. If not, please wire charges, condition, route, arrival, etc." Someone from Louisville, Kentucky, perhaps trying to save Western Union fees, wrote simply, "Please send penguin good home love pets wire collect."[12] Seemingly, everyone was interested. Children, housewives, lawyers, veterans, teachers, farmers, zoo directors, and even US senator William D. Smith.

Most people who were concerned about shipping assumed that the penguin, if they got it, would be sent by mail. One remarkable telegram, however, offered up another suggestion, one that all penguins may yearn for. The principal of the Bruce School, Kansas, Missouri, was so certain that he would get a bird that he had "arranged with TWA [Trans-World Airlines] for transportation by air of one or two penguins."[13] The possibilities seemed endless.

Clearly not understanding the rarity of penguins or the difficulty in getting them safely to America, one writer from Tennessee added a postscript to his penguin request: "If all are gone, please bring some more back from the Antarctic."[14]

PENGUINS, DOA

Penguins are not, in fact, easy to transport. Though they look solid, they are birds sensitive to changes in environment. Wild animals, they are unused to being fed dead fish, or being fed by humans at all. Often early attempts seem to have failed because the birds simply wasted away. Finally, penguins who are moved from one habitat to another often suffer from mycosis, a respiratory illness that was determined to be the leading cause of zoo deaths in the birds by the 1960s.[15] Byrd, having struggled to acquire, transport, and maintain penguins at several points, knew this fragility well. In a letter to the Ringling Brothers circus, he pointed out "P.S. I think I had better call your attention to the fact that acclimated penguins are far more valuable than those that are not acclimated. For some reason or other, certain penguins live very healthily up here and certain others do not. It is about fifty-fifty."[16]

The first zoo in the world to host king penguins was the Edinburgh Zoo, which accepted three penguins from a whaling expedition in January 1913. The zoo, subsequently famous for its breeding program, penguin parade, and for Sir Nils Olav, a penguin knight of Norway, hatched its first chick in 1919.[17] By 1934 Regents Park Zoo had a special, avant-garde Bauhaus penguin pool, which has been preserved because of its architectural value.[18] Norway has another odd moment in the history of importing penguins: in 1936 and in 1938 penguins were introduced to the Norwegian north as "part of a coherent program of using polar animals to naturalize Norwegian authority over polar spaces."[19] As with the Edinburgh Zoo's penguins, these birds were moved via whaling vessels. The last wild penguin sighting in Norway was in 1953.

Though I have not yet fully established a zoo timeline for penguin exhibits in the United States, letters to Admiral Byrd suggest that by 1939 there were numerous zoos across the nation that had acquired the birds. The former Swope Park Zoo (now simply the Kansas City Zoo) had penguins on exhibit; one local remarked in a letter that "[Swope?] Park Zoo in Kansas city has penguins," "and every time we go out there we always go visit them because they always seem so sad."[20] Young siblings in Culpeper, Virginia, had the opportunity to see the birds regularly at "the 'zoo'" (likely the National Zoo).[21]

Even many large zoos didn't get penguins until the 1950s or 1960s. The Oregon Zoo got its first penguins after the director launched and led an expedition to acquire them in 1962. The National Zoo, which no longer houses penguins, similarly sent a zookeeper on a penguin-finding expedition from 1947 to 1948.

Penguins of all stripes were hard to come by and care for, but the famed Antarctic emperor penguin proved incredibly elusive. Robert F. Scott had brought back an emperor penguin egg in 1902.[22] Byrd himself had worked to acquire a living emperor since his first Antarctic expedition. On that trip, he placed Paul Stiple, a young Boy Scout and the winner of a contest, in charge of taxidermy and the sled dogs, hoping that they could also return with live birds. Unfortunately, none of the emperors survived the voyage home. In part, this was because of a lack of preparation; they had no preserved fish with which to feed the birds and were unable to catch any near Little America.[23]

On the second expedition in 1935, Siple, by then trusted with leading smaller parties, was able to keep ten birds alive.[24] He recounted the successful feat in detail, showing not only how much work went into preparing for transporting the birds, but also how reliant the whole venture was on modern transportation and refrigeration technology:

> On leaving the Antarctic we had twenty-one captive Adélies and nineteen Emperors housed amidships in an air-conditioned, refrigerated, cork-insulated room forty feet long, six feet wide, and seven high. Half its length was given over to a concrete tank of cooled sea water three feet deep. Air was pumped down from the masthead and cooled by passing through a large honeycomb coil of ammonia veins. On the cooler days the birds were taken out to a canopy-covered enclosure on deck for a chance to dry their feathers, since the necessity of continually washing down the room kept it too humid and the penguins were sometimes unable properly to dry themselves. After three weeks at sea, when the birds had been kept for two or three months in captivity, most of the Emperors no longer required forcible feeding.[25]

On February 7, 1935, the *Chicago Daily Tribune* reported the imminent return of the SS *Jacob Ruppert*, which sent the following dispatch: "At this writing we are proceeding westward along the front of the tall white cliffs of the Ross ice barrier, bound for Discovery inlet, where a brief stop will be made to collect live penguins."[26] The penguins were to have been one of the crowning achievements of the expedition, as the collected birds included living emperors. Up to that point, no emperor penguins had made it to America alive.

These penguins were sent to the Brookfield Zoo, run by the Chicago Zoological Society. The arrival of the birds was touted in an article from the *Chicago*

FIGURE 8.2. A child's drawing of a penguin in formal wear. *Source*: Record Group 56.1, Series 1, Box 19, folder 740, Byrd Papers.

Daily Tribune. Marcia Winn described the arrival of the penguins in a manner reminiscent of a high society gala: "A strange group of gentlemen will be ushered into the Chicago Zoological park at Brookfield next Wednesday," she wrote, "after their arrival by train from Boston." She noted their starched formal attire and the fancy dinner of "salt and smelts" they would enjoy. Finally, she revealed her joke, since "this fastidious troupe will be made up of thirteen penguins, the entire collection of these antarctic birds brought back from Little America by Admiral Richard E. Byrd."[27]

Unfortunately they did not bring the hoped for attention to the zoo or to Byrd's future expeditions, since they quite quickly died of mycosis. A newspaper account related the death of "another of the Emperor penguins brought back from the Antarctic by Rear Admiral Richard E. Byrd." By this point, the paper notes, only one survived. The surviving bird, the zoo director claimed, "would probably continue to live." The lone emperor was named Lena, and "appears fresh and frisky in the specially refrigerated quarters she is now ocupying [sic]."[28] Unfortunately, that would not prove to be the case, and Lena died as well. The penguins were then taxidermied by the staff at the Field Museum (the former name of the Brookfield Zoo). The "habitat" opened to the public on July 7, 1936, with the penguins "mounted in lifelike attitudes amid a reconstruction of their natural environment in 'Little America.'"[29] These Antarctic birds were visible to many Chicagoans from 1935 through the 1939 World's Fair—first at the zoo and then at the museum, where the birds are still on display. This set of emperors provides both reminder of the fragility of these birds and evidence of the ways in which Little America and Byrd became synonymous with penguins in America during the 1930s.

FIGURE 8.3. Emperor penguin display at the Field Museum. Photograph by Ellen Arnold.

Little America was the name given to a series of Byrd's base camps on the Ross Ice Shelf. These bases were home to scientific endeavors and became an extension of ideas about exploration, American imperialism, and American success. The camps captured the public's imagination. One of the "penguin request" letters underscores this connection. Seventeen-year old Thelma Nurse of Rochester, New York, wrote "I have been very much interested in your broadcasts to and from little America on your last expedition and am looking forward to hearing the ones transmitted on this one." Henry Burch of Maryland asked to be sent a penguin, noting that beyond the general joy of having such a pet, "coming from Admiral Byrd, I shall treasure it most highly."[30] A mother from Ohio asked for penguins for her children, who had "followed your expeditions and breath-taking successes for many years via map and chart, the printed page, the radio and motion picture." She noted some ambivalence about the necessity of further trips, but nonetheless "look[ed] forward with keen anticipation to your rollicking South Pole broadcasts again." She recognized that the penguins may, in fact, be long gone, but concluded her letter by noting that "even a feather from one of Admiral Byrd's penguins would be cherished by my youngsters."[31]

PENGUIN ISLAND

The 1939 World's Fair in New York included Admiral Byrd's Penguin Island, an exhibit that contained equipment used in the second Antarctic expedition, an exact replica of the cabin that he had lived in, as narrated in his popular memoir *Alone*, and other artifacts from Little America. It also housed his husky sled dogs and live penguins.

Byrd had a predictably hard time acquiring the penguins for the exhibit—and keeping them alive. Correspondence between Byrd and Colonel Leo MacDonald, the man in charge of Penguin Island, reveal the admiral's anxiety about the speed of the arrival of the penguins, which had been purchased from the Peruvian embassy via an unknown German intermediary (referred to only as "the German"). Byrd ordered seventy-five penguins, knowing that many of them would die. He wrote to MacDonald, "We should be able to get fifty of them back to the Fair alive." He pointed out in an earlier letter, "Every time a penguin dies I lose a hundred dollars."[32]

Eventually Byrd secured enough penguins, but the delivery time kept being delayed, and he worried that the birds would not arrive in time for the opening. They were scheduled to arrive from Panama by steamship, but the boat was delayed. Byrd hurriedly tried to plan for at least one penguin to put in a punctual appearance, writing to Juan Trippe, the president of Pan American Airways to see if they could fly a bird up from Miami on one of the line's Clipper planes. He wrote, "It will be a bit humorous to have nonflying birds making a date via the air. Hope you can grant me this courtesy." Trippe's response, if any, is not preserved.[33] Eventually, Byrd had to settle for the delayed steamship, and he tracked its progress with anxiety. A *New York Times* article from May 2, "Byrd Lands 119 Penguins" presented the results: 119 penguins arrived at the Brooklyn pier. However, there had originally been 200 birds; 81 had died crossing the equator. With Byrd's characteristic zeal for micromanaging and flair for the press-worthy moment, the piece reports, "Rear Admiral Richard E. Byrd was at the pier to supervise transfer of the penguins to trucks for conveyance to the World's Fair. They are to appear in the admiral's 'Penguin Island' and it is understood effort will be made to teach them tricks."[34]

The penguins at the New York exhibition were from Tierra Del Fuego. One newspaper report noted that the penguins needed new homes because, "Admiral Byrd can't take them back to the Antarctic, even if he wanted to and they wanted to go, because they didn't come from there and probably would quickly die."[35] This is one of the few acknowledgments of the mis-placing of these penguins. Byrd seems to have been ready to use any penguin in his Antarctic show, despite his determination that his own workstation be very precisely reproduced. Penguins, were already, it seemed, interchangeable. Yet the newspaper author was quick to point out that the birds had thrived in New York, contributing to the assumptions of Byrd and others that the penguins' home environments were not necessary for their thriving, and that America could be a very fine place for penguins. He reported that the birds "did so well [at the Fair]—being admired daily, that they gained about seven pounds apiece."[36]

People had heard Byrd speak in local theaters, watched newsreels, and even, in the case of fifteen-year-old Eleanor Wilson of Blacklick, Ohio, toured one of Byrd's ships and received an autograph.[37] A Missouri teen had also met Byrd, telling him that "I went to see you two years ago went you were at Springfield, MO. I injoyed hearing you and seeing your movies their [sic]."[38] Yet most people who wrote in for one of the World's Fair penguins had never encountered either Byrd or the birds. They depended on modern technology to mediate their encounters with penguins. A young man in Pennsylvania wrote, "I always wanted to have one ever since I saw them in the movies and the way they act."[39] Film, photography, and radio all exposed people widely to Byrd's expeditions and adventures, helping cement his celebrity status.

Perhaps the most ubiquitous and widespread way that Americans learned about Byrd and the Antarctic was through a series of weekly broadcasts from the second Little America base camp. From 1933 to 1935 CBS and General Foods sponsored regular radio broadcasts from the camp. In his article on the broadcast Stephen Perry notes that "though rarely noted in radio history literature and nearly lost to time," because they were not recorded, the broadcasts are important because of the financial sponsorship and "product placement" of sponsors and "because of its role in allowing the average citizen to explore vicariously to the ends of the earth."[40] The program helped to both promote Byrd and his sponsors and to prove the vibrancy and reach of radio technology.[41] The show contained news, updates, and conversations with the explorers and updates on entertainment news. There were also music pieces composed especially for the program, including a piano solo called "Penguin Cakewalk." The episode from December 5, 1934, was devoted entirely to penguin research.[42]

PENGUINS, COD?

Why did so many people imagine that a penguin could simply be mailed to them? Some answers may lie in the rapid expansion of the role of the post office in the decades leading up to 1939. As Danny Lewis writes for the *Smithsonian*

Magazine, "When the Post Office's Parcel Post officially began on January 1, 1913, the new service suddenly allowed millions of Americans great access to all kinds of goods and services." COD (collect on delivery) was set up six months after the start of parcel post.[43] The ability to mail awkwardly sized parcels was a remarkable service, helping to deepen connections between communities across the nation and erase some of the disparities of access to goods that rural communities faced. And though some parcel delivery was still done by horse-drawn carriage in the early days, the railroad quickly became a partner in regular "speedy deliveries."

When parcel post was established, a new world of possibilities opened for people to exchange goods. According to Nancy A. Pope at the National Postal Museum, people began almost immediately to ship a startling variety of goods: eggs, pitchforks, silver spoons, prunes, pancake flour, brooms, baked beans, and even a roast chicken. Perhaps most famously, some people even mailed children. The first recorded incident, widely reported at the time, was an Ohio couple who mailed their eight-month-old son to his grandmother (who lived nearby) for fifteen cents. This practice was short-lived but widely publicized (the last known incident was in 1915), perhaps priming the pump for the idea that parcel post could be used to transfer other living creatures.[44]

But there were, after all, some limits. Nancy Pope notes that "among the items considered unmailable in the new service were intoxicants, poisons, poisonous animals, insects, reptiles, inflammable materials, pistols, revolvers, live (or dead) animals, live birds or poultry, raw hides or 'any article with a bad odor.'"[45] It does therefore seem surprising that so many people thought that they could receive a penguin through the mail, COD. There are two contributing factors that may help us understand this. The first is the presence of a post office at, Little America. The second is the popular children's book *Mr. Popper's Penguins* (1938).

On October 6, 1933, the US Postal Service officially opened a postal station in Antarctica, located at Byrd's Little America base camp.[46] (The camp, on the Ross Ice Shelf, is now lost as it was located along one of the fault lines where the shelf broke away.) The establishment of the post office worked as a vehicle for drawing support and attention to the exploration of Antarctica, and was used chiefly to mark and cancel stamps issued in relation to the expedition. To get a canceled stamp sent from the South Pole (or close enough), Americans across the nation merely had to send in 53 cents (50 cents of which went directly to fund the expedition). The postmaster reported that over 150,000 pieces of mail were canceled at the station, many of them were related to stamp collecting. The station was open only as long as the expedition lasted, but clearly the connection was made in the American imagination between the US mail and polar exploration. In fact, Little America's second cancellation stamps featured penguins carrying mail pouches. Postal penguins were circulating by 1939.[47]

With best wishes
Paul A. Siple

U.S. MAIL OFFICIAL CACHET

(C B a n II

BYRD ANTARCTIC EXPEDITION II
LITTLE AMERICA, ANTARCTICA.
SECOND CANCELLATION
MAIL

LITTLE AMERICA
JAN 30
12 - M
1935
ANTARCTICA

BYRD ANTARCTIC EXPEDITION II
3 U.S. POSTAGE CENTS 3

BYRD ANTARCTIC EXPEDITION II
3 U.S. POSTAGE CENTS 3

Oscar F. Cartwright
Entomology Division
Clemson College
South Carolina .

FIGURE 8.4. A "cachet" stamped at the Little America post office, digitized by the Smithsonian and available as a copyright free image on their site under "Antarctic Post Office." *Source:* "Object Spotlight: Antarctic Post Office," National Postal Museum, https://postalmuseum.si.edu/collections/object-spotlight/antarctic-po.html.

We also need to consider the option that a young fan of Admiral Byrd pointed out, that many people had already read the funny and popular *Mr. Popper's Penguins*. The book had a plot so eerily like the real events playing themselves out in November 1939 that it could almost have been a publisher's publicity stunt. In the book a regular working-class couple is listening to a radio broadcast of the famous polar explorer, Admiral Drake, clearly modeled after the broadcasts from Little America. Unexpectedly Drake greets Mr. Popper by name, thanking him for his letter, and adding, "Watch for an answer. But not by letter, Mr. Popper. Watch for a surprise." The next day, the doorbell rings, and it is "an expressman with the largest box Mr. Popper had ever seen." The package came "Air Express all the way from Antarctica," and was covered with labels and instructions, including "KEEP COOL." It had holes punched in it and (unrealistically and dangerously) a layer of dry ice. It was, unsurprising for us but definitely for Mr. Popper, a penguin.[48] Hilarity ensued.

Finally, it bears remembering that zoos did ship animals across long distances, though the average person writing to Byrd would have had no direct knowledge of this. The records for the Board of Directors at the St. Louis Zoological Gardens shine some unexpected light on practices of penguin shipping. On June 15, 1928, the board was requested to approve "the following express charges on animals," which included "two Sea Elephants, twenty-seven Penguins, nine snakes, 2 Ringtail Lemurs, from J. T. Benson, Nashua, N.H.," costing a total of $456.36. Those same penguins had cost $4,625, with two being donations, and the sea elephant had cost $7,500, and so once math magic happens we can guestimate that it cost the zoo approximately $7 to ship a penguin express in 1928.[49]

Whatever the constellation of factors, it is clear that the response was overwhelming. Most applicants assumed the birds were free. The Admiral's staff quickly issued a correction, explaining that the birds were for sale and indeed quite dear, but not before a deluge of mail, telegrams, and in-person requests arrived. The *Chicago Daily Tribune* for November 7 reports that the deluge of penguin requests delayed the trip by a day. It describes a scene that must have flustered even Byrd's resolute and practiced staff: "At his home, at his office, and at the navy yard dock, where the two expedition ships were loading, virtually nothing could be done today except turn away applicants." The reporter added that "some optimistic applicants arrived early, at the admiral's home, in two trucks, ready to carry off as many birds as possible, altho the birds are at a New York aquarium."[50]

THE PERKY PENGUINS

This chaos was clearly felt at Byrd's offices; one person in particular, Byrd's secretary Hazel McKercher, seems to have borne the brunt of this unexpected penguin frenzy. Though all records do not survive of her end of this correspondence, she was still writing letters back to the public in February of the following year. For example, in response to sixteen-year-old Joan Rigby's letter of November 6, McKercher replied, "I greatly regret the delay in replying to your letter of inquiry about penguins but it was just on the eve of Admiral Byrd's departure for the Antarctic and we were deluged with hundreds and hundreds of letters." She noted the mistake in the AP bulletin and added that the birds "were very comfortably housed in an aquarium and well taken care of." She concluded by noting that some were sold for $150, since, "as you probably know, penguins are the rarest birds on this continent and we secured them at great cost, and we could not afford to give them away."[51] Indeed, in 1928 the St. Louis Zoo paid $185 apiece for twenty-five penguins. Costs seem to have fluctuated based on the species's rarity: in 1927 they paid $750 for twelve Humboldts; in 1928 they paid $150 each for four Magellanics, $200 each for six rockhoppers, and $100 for a black-footed penguin.[52]

Fearing a loss of significant income following the financially devastating exhibit, Byrd tried reaching out to all manner of connections far and wide to find buyers for his twenty-plus penguins (despite newspaper reports, the number of actual penguins is never entirely clear—internal correspondence from Byrd's team suggests that there were either twenty-four or twenty-five birds of various species). It appears that MacDonald had been working for a while to find buyers for the birds at $150 apiece, with few results. He first tried to sell them to the Ringling Brothers circus, writing on October 16:

I have got 25 penguins which I have had for six months and which have become thoroughly acclimated. They are healthy birds. I should think they would be a splendid thing for your circus. If you could have a glass tank made (it need not be a big one) you would have an extraordinary attraction, as penguins swim beautifully, faster, in fact, than a fish. They go like a streak through the water and people never tire watching them. The children are awfully keen about them, and I believe the advertising of these penguins would increase considerably attendance at your circus.[53]

Henry Ringling Worth countered Byrd's initial price, offering only $50 per bird. Frustrated, the admiral wrote to MacDonald on November 2, "The circus would give us only $50.00 a penguin so that's that. I feel I would almost rather give them away than accept any such amount." The next day he wrote to MacDonald again, frustrated: "We are caught with the penguins on our hands just as I thought we would be."[54] And here the problems seem to have begun: Byrd tried to micromanage the sales, eventually deciding to try to get a reporter involved, the results of which we have already seen.

McKercher wound up having to triangulate between Byrd, MacDonald, and the director of the New York Aquarium (where the penguins were being held), Dr. C. W. Coates, as they tried to sell the penguins (for $150–200 apiece) to zoos, aquariums, or wealthy patrons.[55] She followed up on any letters or telegrams that seemed promising, including ones from directors of the city parks in St. Paul, Minnesota, and Rochester, New York, who had expressed interest in the penguins for zoos. MacDonald even tested out a scheme of renting the birds for $100 a week (including a trained attendant).

Though I cannot trace the final disposition of all twenty-four (or twenty-five) birds, most are accounted for. One died. McKercher sold eleven penguins, including two to Colonel William Horlick (of the Horlicks syrups company) in Racine, Wisconsin. MacDonald sold three penguins to an unknown buyer (at least to McKercher, who was keeping the records). There were also a series of "partial" or potential sales, including to the Marine Studios at the American Museum of Natural History. By December 1940 seven penguins were still in the aquarium.

McKercher also handled inquiries on the care and housing of penguins, dealing with "how should they be shipped to St. Paul, Minn., any special care enroute, etc. even to the point if they are good playmates for four year olds." The birds were tricky to care for, and both MacDonald and Coates wrote up careful instructions, which McKercher then had to type up and send out to all of the penguin purchasers. Coates acknowledged the burden on his staff as well, telling McKercher that "we have written so many letters describing the needs of these birds that my secretary considers them as almost a form letter."[56] After taking a Christmas holiday, McKercher wrote to Coates, relieved, "I am glad to have [the penguins] off my hands although they are rather amusing birds."[57]

In the records of the sales, we do start to finally answer the question about the shipment and transportation of the penguins. Two were bought by private citizens, the Haffenreffer brothers, who donated them to "the children of Rhode Island for Christmas and to be kept at Roger Williams Park." These birds were picked up in person by the superintendent of parks. E. J. McGuire of St. Paul, Minnesota, also purchased a pair of penguins "to show at the St Paul winter Carnival.[58] These birds were indeed sent by express mail. McKercher wrote, "I called the express company here and they tell me that live birds are shipped at 'a rate and a half' and the charges on a 50 lbs. shipment (I estimated something for a crate) would be 4.74. It might be a little less from New York." We also learn that the Humboldt penguins took to relocation relatively well: "They were so tame," Coates noted, "that our own keepers could take them out of the shipping cage immediately after a long imprisonment hauling by truck without the slightest danger." Whereas the Humboldts were "well-behaved creatures," the remaining black-footed penguins, "against which everybody should be warned," did not enjoy being shipped, or touched, or tamed.[59]

Coates kept records of the shipment of several penguins and wrote on February 5, 1940, that he "shipped these birds (three) to [Mr. Grace] on the 4.05 pm train today, to arrive early Wednesday morning." He then explained that "our practice is to send such shipments 'express collect.'" Along with this letter he included a statement of the "various costs incurred in the shipment of these various birds." The shipping fees totaled $19.69, which included crates, telegrams, and the delivery of the birds to the trains by taxi! He explained that the costs were variable, depending on distance, size of crate, and number of birds. At the point of writing, the aquarium was still caring for those seven birds, one which was unfit for sale, and so he reminded McKercher, "You are still in the penguin business."[60]

Most notable of the private collectors was Evelyn Walsh McLean (the owner of the Hope Diamond) who purchased two penguins for $350 and tried to keep them in her daughter-in-law's Midtown Manhattan apartment. A newspaper clipping notes that "the hotel carpenter built a set of penguin stairs so they could climb in and out of the bathtub with ease." Yet after one evening, "the birds were sent back to the Aquarium until they could be taken to the summer home in New Jersey and then migrate around the US with their wealthy owners." "Later in the season," the article continues, the birds were to fulfill Byrd's imagined trek and "go to Florida by plane."[61]

As they were transported out of their natural environments and their communities, the birds became homeless in multiple ways. Their displacement, combined with the fact that people could see these birds "at home," also may have allowed people to imagine having a personal relationship with a particular

FIGURE 8.5. Pete the Penguin at Youngstown State. *Source*: Courtesy of YSU Archives and Special Collections, William F. Maag, Jr. Library, Youngstown State University, Youngstown, Ohio.

penguin and being able to have one in their own homes. This was demonstrably a shared vision of the hundreds of people who sent in letters, but it was only possible for the very rich and well connected to achieve.

The story of one of these penguins encapsulates the whole affair. McKercher sold one of the Humboldts to Howard W. Jones, the president of Youngstown State College in Ohio, where it lived briefly as the school mascot, "Pete the Penguin." The historical summary of Pete or Petey does note his Worlds' Fair origins, explaining how he wound up at the college: "John Chase, Youngstown naturalist and geologist, noticed newspaper announcements which said that Admiral Richard E. Byrd was planning to dispose of several penguins brought back from his last trip to Antarctica and for the past year resting at Penguin Island at the New York Worlds Fair." Of course we already know that this was a partly mythical origin story, as these penguins had been purchased specifically for the fair. Petey's established narrative continues: "Apparently there were plenty of animal lovers ready to take a penguin for a pet, but none of them ready to pay the cost—$150.00." "After considerable writing, wiring, and telephoning," the college was told that they could purchase one of the penguins, and that it would be "shipped immediately." The college received the care instructions that McKercher had compiled and were a bit daunted, wondering, it is said, if maybe they should have "secured a more conventional and less unique mascot—maybe a Persian kitty or a wire-haired terrier."[62]

The story of Pete's arrival appears to have gone down in college lore, as the school was hoping that he would arrive in time for a debut at the school's homecoming game. Officials waited anxiously at the station for the train from New York City, but no penguin arrived. Disappointed fans were told "over the broadcasting system," that the penguin would not arrive that day. The game was already well into the third quarter, however, when "suddenly, two excited expressmen carrying a badly shaken, crated penguin ran into the stadium. Yes, the game was stopped while officials, payers, and the homecoming crowd got a look at the first mascot of Youngstown college."[63]

Bringing this story of mass communication, modern transportation technology, penguin popularity, and Byrd's birds to a fitting conclusion, Marilyn Chuey, the daughter of President Jones claimed that "it is believed that Pete was the first penguin ever to be interviewed over the radio." Pete, however, was silent. For her thesis Chuey interviewed a reporter who had been there, and who observed that it wouldn't have mattered if Pete had been vocal anyway, as "not a person among the thousands in the listening audience would have understood one quack. Though other members of Admiral Byrd's flock of penguins were scattered all over the United States, it was not likely that any of these were listening in."[64]

There are moments in the Byrd correspondence when experts privately and to one another acknowledge species differences and address the struggles of adapting penguins to both northern climes and captivity. However, across the press and public responses, and in Byrd's public statements, people seem ready to imagine penguins as both safe and *happy* as displaced animals. Despite increasing awareness that penguins lived communally in the wild, people also often encountered and imagined them as individual birds. As Eileen Diamond of Kingston, New York, pointed out, "I can't possibly take care of twenty but I'm sure I could get along with one."[65]

Finally, in the many appeals by private citizens to have their own penguin, and in the story of Pete in Ohio, we also see the start of a new kind of discourse around penguins—that they had a type of individual character with whom people could form new relationships. William Grubb (age twenty-four) of Milwaukee explained, "I don't want it for a show off or to put in a cage and forget it, But I want it for a pal"; and Doris Klutz of Galveson, Texas, who had to go to a convent school because of a sick mother, wrote, "I love little birds and animals to play with. I am at home all by myself I have no one to play with so please send me one."[66]

In modern America, it seems, penguins could be both interchangeable in the aggregate (any penguin could stand in for all penguins) and be ascribed individual personas, knowable life histories, and agency (for example, as a radio guest, or a pet, or a friend). This may have to do with the mediation of technology. As the editors point out in their introduction, media brought the wild "home." The more exposure to penguins people got, either in person in zoos or via the new film reels, through novels or over the radio, or as toys and trinkets, the closer it allowed them to imagine being to having penguins in their homes.

CHAPTER 9

SEEING THROUGH WINDOWS, WALLS, AND THE WEB IN PUBLIC AQUARIUMS

DOLLY JØRGENSEN

Our human bodies are limited to certain environments—we must breathe oxygen from the air and our cells cannot withstand high pressure. Other than in very shallow water, humans need technologies to interface with water-based creatures. This means that our knowledge of aquatic life has been highly mediated by technology. The only firsthand contact the vast majority of people will ever have with ocean life is in an aquarium. The first public aquarium opened its doors in 1853 at the London Zoo, and now there are large public aquariums across the world. From the beginning public aquariums were venues to display marine and freshwater aquatic species, as well as reptiles, birds, and marine mammals, to the public. Unlike your home aquarium, which is typically a small self-contained unit, public aquariums are highly complex systems.

Public aquariums, as planned and constructed spaces for this human encounter with the otherwise unreachable animal, are technologically mediated landscapes. There are material realities behind the experience associated with them. From a technology perspective, pumps and electricity are primary technological drivers for what we see them today: the water has to be changed and circulated; salinity and temperature has to be regulated; the lighting has to be matched to the species and visitor needs.[1] All of those things become possible only once you have electricity that can be supplied to the aquarium. There are also skills—"tank craft" as Samantha Muka has called it—needed to build and maintain tanks that are livable environments for undersea life.[2] One material reality in the aquarium experience that has not been significantly or historically explored is the container—the glass through which we see undersea life. The technology of the container matters deeply because of the changes it creates to the visual experience of visitors and thus their closeness to nature.

Most of the interaction between human and animal in the public aquarium takes place through sight. Although touch tanks allow visitors to feel the skin, spikes, or shells of animals and sounds are often used to create immersive underwater experiences, seeing life underwater is the driver for aquarium design. Humans are typically ocular creatures. Although hearing, taste, touch, and smell shape our daily lives, seeing has been considered our leading sense since Aristotle.[3] The eye became a symbol of the intellect and was considered the most objective. Although there are humans who do not see and thus rely on other senses to make sense of the world, the privilege and priority of sight has continued into our media age as we "watch television" or "see a film," even though hearing is just as involved in the experience.[4] The visual nature of the internet, whether those visuals come in textual or pictorial form, has dominated its design and usage to date. The technologies of seeing, especially in the liquid environment that is so inaccessible to humans without help, matter in the interaction between humans and aquatic animals.

In this chapter I am particularly interested in exploring how technologies mediate our experience of seeing animals.[5] Mediation has been an important strand of thinking in Science and Technology Studies to understand how technologies shape human lives. Mediators "transform, translate, distort, and modify the meaning or the elements they are supposed to carry."[6] Thus the transformation, distortion, or modification caused by a mediator can affect both the content itself and the perception of that content. Glass as a medium is supposed to be transparent and unseen, yet it also distorts, magnifies, and changes the angle of the view. In the aquarium the materials (glass and its later replacements) involved to make visualization possible have a profound effect on both the contents of the aquarium and the way they are viewed. A focus on the interface of the seeing allows us to see the mediation of these media. The glass in this way moves from unseen to seen.

There has been a surge of recent scholarly interest in how underwater landscapes are aestheticized and the role of scientific inquiry into envisioning the depths of the seas.[7] Scholars such as an Ann Elias and Jonathan Cristopher Crylen have shown how underwater photography and filmmaking depended on innovative ways to separate the camera's lens from the aquatic environment, including John Ernest Williamson's photosphere and Frank Hurley's on-land aquariums, that set later aesthetic expectations of what the undersea realm was supposed to look like.[8] Media and theater scholars have moved to incorporate aquariums, as well as underwater photography more generally, into media history. As Massimiliano Gaudiosi noted, aquarium practice has "a technological component and a repertoire of representational strategies" influenced by as well as influencing cinematic practice.[9] In her study of the home aquarium, scholar of theater and dance Judith Hamera has argued that aquariums deployed the "spectatorial technologies" of the window, theater panorama, and diorama.[10] This scholarship has either focused on the period of early aquariums or stayed within the context of small home aquariums, which has tended to limit discussion of technological change in aquarium presentations.

This chapter will examine how public aquariums have undergone transitions in the way they present natural knowledge of undersea environments because of technological change in available materials. I will place each of the modern aquarium's technological mediation windows—glass, acrylic, and digital—into a longer historical context. By paying attention to the interface, I want to show how the technologically mediated undersea experience in the public aquarium creates new connections between human viewers and underwater creatures.

WINDOWS OF GLASS AND SEEING SIDEWAYS

Humans have been looking at fish for millennia. Fish ponds built for aesthetics are common features of ancient civilizations, from Chinese ornamental water gardens to show off goldfish to Roman *piscinae* for both saltwater and freshwater species. By the tenth century Chinese aristocrats had placed their prized domestic goldfish temporarily in ceramic fish bowls in the house to show them off to guests; after the viewing, the fish would be returned to the koi pond. Until the nineteenth century the mode of observing fish was from the top, with the viewer standing over the pond or the bowl looking down at the fish.[11]

The glass aquarium changed this mode of seeing. The glass terrarium (known then as the Wardian case) was invented by Nathaniel Bagshaw Ward in 1829 to seal plants off from the outside environment and was quickly adopted to transport living plants from across the globe to botanical collectors in Europe.[12] Because Wardian cases depended on air-tight seals, they were soon adapted to reproducing aquatic environments. By 1850 articles started appearing on how to keep aquatic life alive in these cases without flowing water.[13]

Key to the development of the aquarium was the glass. The glass served as a transparent medium for letting both light and the human gaze into the water. Rather than looking down on the fish from above—a position that creates light reflection on the water, and the fish are difficult to distinguish—fish could be seen from the side swimming within their natural watery habitat. As Susan Davis remarked in her study of Sea World parks, "Aquariums are little experiments in making the invisible visible."[14] While this is a true observation when the focus is on the sea life, an aquarium also attempts to make the water and the glass invisible.[15] Intimacy with the underwater inhabitants of an aquarium was possible by bringing the aquatic life onto the shore in a space that a human could be close to and could see through.

The Fish House at London Zoological Gardens opened its doors in 1853 as the first public aquarium. A collection of marine ringed worms and sea anemones amassed by one of the earliest aquarium experimenters, Philip Henry Gosse, formed the core of the display. Gosse then added fish, mollusks, and crustaceans to some of the tanks at the request of the London Zoological Society.[16] H. Noel Humphreys, a contemporary commentator who wrote the first history of aquariums, noted that these tanks gave visitors "a glimpse of the wonders of the 'ocean floor.'"[17] This view reversed the land-based gaze of humans: "On the land, we have, as the ordinary aspect of Nature, the green herbaceous mantle of the earth below the eye and the azure sky above; while a spectator, standing beneath the water on the ocean floor, would see these features more than reversed: he would see above him a liquid atmosphere of green, and below, an herbage of red or of purple hue, exhibiting strange yet exquisite forms, such as no terrestrial vegetation displays."[18]

Most of the displays in the Fish House were a series of independent small tanks. The sides of the tanks were not glass; instead metal was used. This had to do with the water pressure and how of it much glass could withstand with the construction technologies they had at the time. The design was also linked aesthetically to the iron-and-glass aesthetics of Victorian greenhouses, which was the contextual genesis of the Wardian case. Controlling light through the glass was of particular concern; light from only the top rather than the sides created the most desirable aesthetics.[19] The tanks on the walls were constructed with large plate glass up to 6 feet across, but they were not completely filled with water. They displayed coastal habitats that had only a portion of the habitat submerged.

An aquarium craze quickly spread in Europe. Publications on how to create home aquariums rapidly entered the market, including Philip Henry Gosse's *The Aquarium: An Unveiling of the Wonders of the Deep Sea* (1854) and Shirley Hibberd's *The Book of the Aquarium and Water Cabinet* (1856).[20] When the Jardin Zoologique d'Acclimatation in Paris opened in 1860, it included an aquarium that housed both fresh- and saltwater animals. A view of the interior, published as an engraving in October 1860, shows large glass tanks in the walls.[21]

The earliest public aquariums in Paris and London were embedded in a zoological garden context. Zoos were the perfect places for aquariums. The modern zoo emerged in the late eighteenth and early nineteenth centuries as "display and experimentation spaces for the scientifically inclined."[22] Animals in these zoos were "organic objects of desire"[23] put on display to reinforce both imperial spectacle and scientific classification of nature.[24] Zoo aquariums displaying underwater creatures that humans would not normally encounter fit the goal of zoos to bring the most exotic and wild species into domestic spheres.

Being able to see undersea life in a tank was critical to the experience. One writer from 1895 commented on the visual design of exhibition aquariums: "In order to obtain the most striking effects, the tanks should be in convenient

FIGURE 9.1. Drawing of Mulertt's Aquarium at the Cincinnati Centennial Exposition, 1888. *Source:* "Exhibition Aquariums," *The Aquarium* 3, no. 36 (July 1895), 186.

height to the eyes of the observer; the shape of the tanks should be long rather than high, with sufficient depth (toward the rear) to admit the formation of a proper background."[25] The tank was also supposed to be well lit, but the area where the viewer stood in front of the tank should be dark so that the viewer's eyes could focus on the tank's interior. The position of the viewing was critical to allow the visitor to be at eye level with the undersea life, seeing things from their perspective. The eye-to-eye view of fish was aesthetically desirable to make them more beautiful.[26]

These aesthetics mirrored changes happening since the 1880s when modern show windows became fashionable. At that time shop windows started appearing with appropriate lighting and care being taken in design, color, and placements of the objects for sale;[27] *The Show Window*, a trade magazine, was started in 1897. The aquarium was a contemporaneous development in terms of creating illusions of desire as through a shop window. A drawing from the Cincinnati Centennial Exposition of 1888 demonstrates the creation of desire with its aquarium exhibition (fig. 9.1). On the right, there is a well-dressed woman visitor staring into a window similar to what one might do when window shopping.

The aesthetics also mimicked the diorama to a certain extent. Art historian Guillaume Le Gall has compared the public aquarium at the Jardin Zoologique

FIGURE 9.2. New York Aquarium interior. *Source:* "The New York Aquarium Opened," *The Aquarium* 4, no. 42 (January 1897), 83.

to Louis Daguerre's diorama theater of the 1820s with its use of transparency, transposed images, and movement to evoke wonder.[28] The natural history diorama, which developed at the end of the 1800s, likewise attempted to evoke wonder through theatricality by bringing a natural scene to life, even though the players on the scene were motionless taxidermy specimens.[29] The diorama focused on putting the animal into its correct habitat (seabirds roosting on rock cliffs or deer in a forest or bison on the Great Plains), often with complex scenery props and painted backdrops. Aquariums in this first phase could do this only to a very limited extent because of the size limitations. While aquarium enthusiasts did favor the inclusion of sessile organisms such as anemones and corals in their displays, there was a limited sense of the aquatic environment as an ecosystem. Instead, aquariums tended to focus on "little jewel tanks holding single species" while glass was still the medium used.[30]

The first big modern, stand-alone (i.e., not connected to an existing zoological garden) aquarium was opened in New York in 1896.[31] On the first floor, its largest glass wall tanks were 7.5 feet (2.25 m) across and smaller wall tanks were 5 feet across, and on the second story there were smaller tanks that were either 3 feet or 5 feet across (fig. 9.2). The aquarium featured four to five thousand fish on display at any given time. Because of their small size, the fish tanks allowed

FIGURE 9.3. Common pike exhibited in the Detroit Bell Isle Aquarium, 1890. *Source*: LC-D470320, Library of Congress, Washington, DC.

for only a limited number of species and individuals in each tank. Which species were shown where tended to change quite frequently based on fish stocking and death, so the guidebooks for the New York Aquarium provided descriptions of the fish possibly on display along with the "fish number," which the viewer could then match to a fish number label placed above the tank.[32] The focus in these type of tanks were individual fish types that would be encountered at eye level in order to observe the fish's aesthetic qualities.

At the front of the aquarium open pools were built for the larger species in the collection, including seals, manatees, and larger fish. For the most part aquarium designers wanted to limit the number of open tanks because visitors tended to either feed the fish (things that fish should not eat) or try to catch them (on more than one occasion young boys with fishing lines were caught inside of the New York Aquarium).[33]

This kind of aquarium design as shop window or even as artwork is also evident in the Detroit Bell Isle Aquarium in 1904. The exhibition room called the Grotto had forty-four wall tanks, three floor tanks, and three large pools.[34] The aesthetics of the aquarium stressed the individual tanks and windows. It is as if there were individual portraits of each fish through each window. The label above the tank gave the name of the fish in the tank, like the title of an artwork.

In fact, each tank was framed with the same kind of wooden decorative frame as would be found around artworks on display (fig. 9.3).

In this early phase of public aquariums, which lasted from its beginnings in the 1850s until the 1970s, they were designed as show windows for objects that had previously been unseen, displaying fish as individual pieces behind individual windows, aesthetically presented in order to be pretty or desirable. Visitors passed through the space aesthetically, like an art gallery in which each small individual work of art can be observed (or skipped over). The limitations of glass as a material limited the size and complexity of public aquarium tanks. The glass itself also mediated the interaction with underwater animals. The only way to get to eye level with the fish was to stare through the glass.

WALLS OF ACRYLIC AND THE AQUARIUM REVOLUTION

Early on everyone recognized that there were problems with glass technology as a mode of seeing fish. Glass creates unequal refraction, which distorts the contents if the glass is thick. This is not a desirable trait if you are trying to design something to be aesthetically appealing. So glass needed to be thin enough to not create too much distortion. But having thin glass was a serious limitation because glass is not a particularly strong material and water is very heavy. As Henry Butler noted in his book *The Family Aquarium* (1858): "If manufactured of any size worth consideration, they will not stand the pressure of the water, and are liable at the slightest touch to burst into fragments."[35] This was actually known to happen; an aquarium with glass walls too thin or too large would burst when someone touched it because the water exerted too much pressure on the glass. The only solution to this was to keep the tanks small and make their sides and bottoms out of slate, marble, or metal, and then fit the whole thing into a solid frame that could take on some of the water pressure.

Not unexpectedly therefore people were eager to find better materials for aquariums. A short article in *DuPont Magazine* in 1923 touted safety glass as a potential solution.[36] The article noted that glass easily broke on aquariums, but if the aquarium wall was made out of safety glass, this would not happen. DuPont's new invention consisted of two sheets of glass with a sheet of polymer in between. In this setup, if one side of the glass breaks, the other side does not; thus the tank stays intact. This was a potential solution to the aquarium glass problem, but it does not appear that it was adopted in the public aquarium milieu. Instead, aquariums ended up turning to an entirely different type of material—acrylic.

Acrylic is a polymer. When it is formed into a sheet, it is often called acrylic glass, although it is not actually glass in its material composition. The first commercial brand of acrylic glass, Plexiglass, was introduced in 1933 by the German company Röhm & Haas as a transparent, hard, and shatterproof replacement for plate glass. It was seventeen times stronger than plate glass and weighed half as much. On top of that, it could be molded into any shape. The primary use of acrylic glass in its early phases was for the military, particularly for the cockpits of airplanes, which required curvature for aerodynamic purposes. Although regular molten glass could not easily be made to fit around a cockpit, acrylic could be poured of a mold into the needed rounded forms; its increased strength and lighter weight was perfect for airplane use.

After World War II a large part of the military market for acrylic glass disappeared. Companies like Röhm & Haas and DuPont, which made Lucite their acrylic glass brand, started searching for alternate objects to make out of acrylic. All kinds of household products entered the fray, and one of them was acrylic aquariums. These were home aquariums, rather than those for large public institutions, and often were sold as novelties in unique shapes.

Acrylic was not however automatically a perfect solution for aquariums. Acrylic panels still needed to be thick to withstand the water pressure in large tanks. In large polymer pours, there was a risk of getting air bubbles into the matrix, which would create visual distortion when used as a viewing surface. There was also an adhesive problem. No matter how strong a panel is, there need to be joints that connect the panel to the frame and other panels. Those joints will need adhesive, and if it is too weak, the joints will leak. Two companies led the technological developments that solved these problems: Ryoko in Japan began supplying large, thick panels for aquariums in 1966, and Reynolds Polymer Technology in the United States joined the market in the 1980s. These two companies still provide almost all acrylic panels to public aquariums today.[37]

The big shift toward acrylic panels in public aquariums received a boost in 1984. A new modern aquarium was under construction in Monterey Bay and the aquarium wanted to make signature exhibit of giant Pacific kelp. Giant kelp plants are extremely tall, reaching from the sea floor up to the surface. So what they needed was an extremely tall exhibit, something not possible with plate glass. The designers decided to install acrylic panels that were 16 feet (slightly under 5 m) tall and 8 feet (2.4 m) wide. These were huge acrylic panels for that time. Multiple panels were attached to frames in order to increase the tank's visibility. Instead of thinking of glass as a horizontal space at eye level, the new acrylic glass would extend the tank dramatically upward. It radically altered the relationship between the visitor and the exhibit and was seen as a way of bringing the visitor entertainment. One of the biologists, Steve Webster, who worked on the Monterey Bay Aquarium development, said, "We realize that it's largely entertainment dollars that are standing in line out there. [We are] making effective use of people's goose bumps."[38] Audiences crowded into such huge and impressive displays.

A public aquarium building boom took off in the wake of Monterey Bay's success. All modern aquariums began to contain exhibits with very large tanks

FIGURE 9.4. The Ocean exhibit of Den Blå Planet, Copenhagen, which opened in 2013. Photograph by Dolly Jørgensen.

observed through acrylic panels. Over time technological developments in acrylics would permit the making of thicker and thicker walls without the visual distortion that would have occurred with glass, enabling larger and larger tanks. The invisibility of acrylic was key to the radically increasing size of public aquarium tanks. The acrylic material was so much stronger than glass that it permitted much larger habitats to be constructed.

Large tanks also had the ability to simulate entire ecosystems, and this became a hallmark of modern public aquarium design.[39] Unlike a small tank with a few fish and plants, the Monterey Bay giant kelp tank permitted showing the entire ecosystem with many kinds of fish, a mix of vegetation, and large structures like big rocks. Tanks encompassing complete ecosystems became commonplace from the 1990s, with expansive views into coral reefs, the Gulf of Mexico, and the deep ocean.[40]

Many large displays started to be fitted with theater-type seating so that visitors could spend more time in front of these signature tanks. For example, the Aquarium of the Americas in New Orleans opened the largest Gulf of Mexico tank in 1990 with theater seating set up for visitors to sit and watch the ecosystem swim past. The increasing size of the tanks permitted the display of sharks, rays, and large pelagic fish at eye level. The improvement in adhesives

FIGURE 9.5. The Seattle Aquarium, which opened its underwater dome in 1977. Photograph by Finn Arne Jørgensen.

has also eliminated the need for metal or concrete framing structures, meaning that panels are invisibly stitched together into extremely large viewing walls that allow the audience to sit and watch the aquatic life swim by (fig. 9.4). In 2014 the largest aquarium window in the world, at 130 feet (39.6 m) across and 27 feet (8.3 m) high, was opened for the whale shark exhibit in Hengqin Ocean Kingdom, China.

When aquariums made the move to acrylic windows, they changed the concept from art gallery or simple show window to theater. In the theater the visitor is invited not to view and move on quickly, but to sit and watch the show taking place underwater. This encourages the visitor to watch for the entrances and exits of animals from the field of vision, observe patterns of movement, and stand close to very large specimens on the move (inevitably small children stand by the wall as in figure 9.4 to get the eye-to-eye view of animals larger than themselves). Massive acrylic walls provide a different interface than either small glass windows or views from above.

At the same time that acrylic panels were being installed as walls holding back water, other acrylic forms started surrounding the visitor. While glass was sufficient for small vertical walls, it is not suitable to hold a great amount of water on top of it. In addition, glass panels generally are flat, meaning that many

FIGURE 9.6. The author's daughter inside a viewing portal within an aquarium tank at Legoland Atlantis, Billund, Denmark. Photograph by Dolly Jørgensen.

FIGURE 9.7. Underwater tunnel in Underwater World, Singapore, which opened in 1991. (It closed in 2016). Photograph by Dolly Jørgensen.

small ones are necessary to create curves. The Seattle Aquarium experimented with allowing the visitors to experience the life of a fish from under the water in 1977 when it built the first underwater dome viewing area, called AquaDome (fig. 9.5). But because of the technological limitations of glass, panes had to be placed within a thick concrete frame, which broke up the view into small segments. The glass panes were positioned to be as vertical as possible, with only one round window in a completely horizontal position at the top of the dome (and thus with the least amount of pressure on top of it).

Acrylic, however, could solve the problem of the small, flat glass panes for underwater viewing areas. The first tunnel in a public aquarium was built in New Zealand in 1985 and the new public aquariums opening in the late 1980s and 1990s typically installed acrylic tunnels. Unlike flat glass, designers realized that acrylic could be shaped into fish-eye shaped portholes, curving walls, ceiling viewing lenses, and tunnels (figs. 9.6 and 9.7). One commentator in 1984 noted the trend toward these nonstandard shapes: "Aquarium designers have steered away from the tired box-in-the-wall approach, opting instead for lively exhibits in dramatically shaped tanks that entice you to peer in."[41]

Aquariums quickly adopted acrylic to create immersion and 3-D underwater experiences for the visitor. In 1993 author Leighton Taylor, in his book *Aquariums: Windows to Nature*, summarized well this development: "Today's acrylic technology provides the next best thing to being wet."[42] Acrylic windows mediate the experience of the aquarium, making possible whole new ways of seeing underwater for the average person by allowing them to be under the water with no special breathing apparatus. Through her own experience of the jellyfish exhibit at Monterrey Bay Aquarium, feminist scholar Eva Hayward has noted that the aquarium space is a "virtual dive" created by "immersing the observer sensually into a heavily mediated space."[43] She notes that the technological conjunction of glass, lighting, sound, and space brings about this immersion. Yet looking through the acrylic (or glass for that matter) is not the same thing as being in the water. While acrylic manufacturers have attempted to make the material as non-distortive as possible, there is still a medium between the viewer and the viewed. The fish, sharks, and rays swimming behind the acrylic wall are in the same building as the viewer, yet they are physically separated in spite of the illusion of coinhabiting space.

FROM WINDOW TO COMPUTER SCREEN

In the digital age aquarium viewing has moved from in front of the tank to in front of the screen—one glass has been replaced by another. There is a question as to how different viewing on a computer screen is from visiting an aquarium in person: while the other senses involved in the physical aquarium visit are removed (smell and sound, in particular), seeing a fish swimming in the glass of a computer screen is virtually the same as watching the fish behind the aquarium glass. In both cases, the glass separates the land-dwelling from the aquatic.

With the growth of the internet in the 1990s, nature-watching became an established digital activity, and one of the first places people started watching were aquariums.[44] A review in 1997 listed a few webcams broadcasting images of aquariums "live," including the Almost Amazing Turtle Cam and Piranha Cam.[45] But the most famous was The Amazing Fish Cam! installed in 1994 in the offices of the web browser company Netscape. It is the longest-running camera website still in existence and has profiles of all the fish that you could watch swimming around the tank.[46] Although this was a new digital phenomenon, watching an aquarium live on a screen actually had a longer history. Norwegian television (NRK) had an in-studio aquarium that they broadcast live to fill small programming gaps in the 1960s and 1970s, a broadcast segment they called *pausefisk* (pause fish). In the early years the tank had Norwegian species, but by the 1970s, it had tropical fish and was broadcast in color.[47] Watching these fish through the TV screen brought them into viewers' living rooms.

Public aquariums also started installing webcams on some of their tanks. Monterrey Bay, for example, has the Coral Reef Cam, Jelly Cam, Kelp Forest Cam, Open Sea Cam, and Shark Cam showing underwater exhibits.[48] The Aquarium of the Pacific in Long Beach, California, broadcasts live webcam feeds of multiple tanks on the explore.org site, including large ecosystems of tropical reef and kelp forest habitats as well as species-specific tanks such as jellyfish.[49] It is difficult to track down when these webcam services were first installed, but sometimes the institution announced the new camera feature in news items. For example, the Birch Aquarium at Scripps in San Diego installed a high-definition webcam in their kelp forest exhibit in 2010, which was touted as providing "a diver's-eye view" of the sea life and a way to "de-stress to the swaying of towering kelp."[50]

Even within the aquarium walls itself, digital viewing has become an integrated part of the visitor experience. In the New England Aquarium, for example, a diver show takes place several times a day in which a scientist enters the Giant Ocean tank in order to perform health monitoring of the animals. During the dive the scientist is equipped with a head camera that permits the in-person viewers to not only watch the diver, but watch what the diver is seeing. When I visited the aquarium in 2015, I watched the Giant Ocean diver show and witnessed how the camera and the underwater view became central to the experience: as the divers with their headcams on were talking to the audience during dive preparations, one mother nudged her child, pointed to the screen, and they both turned to watch the events on the screen for the rest of the dive instead of looking at the diver or the fish through the tank's glass. The merging of screens here is striking—both are similarly showing views behind "glass."

The digital experience of the aquarium mediates the view of the aquari-

FIGURE 9.8. View of the Tropical Reef Habitat in California's Aquarium of the Pacific in person simultaneously with the view online via the webcam during the author's visit. Photograph by Dolly Jørgensen.

um differently than glass or acrylic. First, the webcams can often be placed in the water itself, which reduces the distortion of the acrylic pane and allows for other viewing angles.[51] Second, the light passing through water in an aquarium appears to the human eye more blue than it is because of light refraction and color shift, but a digital image can be color-corrected to show the "actual" color rather than that which the visitor sees in person (fig. 9.8). Webcams can also focus in on small creatures, like the ribbon dragon exhibit at the Aquarium of the Pacific, to show details of sea life.

One main difference between the in-person aquarium visit and the computer-based one is that when the marine inhabitant is in digital form, the human can gaze at it, but it cannot gaze back. Aquatic creatures can look through the glass or acrylic at visitors, and they do react to what visitors do, so the difference with a one-way gaze should not be underestimated. John Berger has argued that humans have adopted a unidirectional gaze toward animals: "Animals are always the observed. The fact that they can observe us has lost all significance."[52] From my experiences watching people interact with animals in public aquariums, this is not true. Observers are always fascinated when the sea life appears to be looking at them and responding to them (whether or not that is actually what the animal is doing). During the COVID-19 lockdowns when aquariums closed their doors to visitors, there were some concerns that some animals were not getting the interaction they were used to having with visitors. With no visitors the Sumida Aquarium in Tokyo organized a Please Remember about Humans event for the public to FaceTime their spotted garden eels so that they would see more human faces—there were two million live views during the three-day event.[53] In this event digital interaction was proposed as a replacement for an in-person encounter on the part of both human and animal.

Media mediate the visual experience. Not only is the view different, but with digital media delivered onto a home or work computer, the fish can be viewed over a long time in the background—there is no rush to get to the next exhibit or keep up with companions. With digital interfaces, the viewer is just as dry as in the aquarium building, but the experience of seeing via the screen is not the same. This does not mean, however, that the digital experience is less authentic. Finn Arne Jørgensen has argued in his analysis of nature webcams that watching animals behind a screen still provides a "real" experience.[54] The live webcam in particular may be a more "authentic" way of viewing nature than documentary wildlife films, which involve significant clipping, repurposing of footage, and narrative creation.[55] As Jonathon Turnbull and colleagues have observed, the "liveness" of digital animal encounters via the webcam requires the human viewer to "be there" to experience it.[56] This leads to a real-life encounter even if it is behind a Windows screen.

THOUGHTS ON THE MEDIATION OF UNDERWATER ENCOUNTERS

Changes in technology has led to shifts in animal encounters in aquariums. The first public aquarium opened its doors in 1853 in the London Zoo. In nineteenth- and most twentieth-century aquariums, visitors viewed sea life in rectangular glass enclosures. The infrastructure created windows into underwater life that would otherwise be inaccessible to visitors. In the 1980s, aquarium design underwent a revolution with the use of transparent acrylic walls. This allowed the creation of huge, multi-window aquarium viewing spaces, including viewing areas that were multiple stories high. Acrylics also allowed the development of immersive environments in the form of tunnels and underwater viewing spaces. In the twenty-first century digital media have moved into the public aquarium and made its experience possible from home on the Web.

My focus on the interface that permits seeing aquatic animals exposes the technologically mediated nature of the human-animal encounter in the aquarium space. As land-based creatures, humans require a technological interface to experience the underwater environment at any significant depth.[57] Technological mediation is required to bring the fish to land and see them eye to eye. There have been changing display media: from small glass tanks to large acrylic surroundings to digital viewing at a distance. As the technology has changed, the interaction style of the visitors and the representing and making of the environment within the tank have changed. The unseen underwater world, normally separated from us because of its foreign medium, is revealed in the modern public aquarium. Yet it is the unseen interface of the glass or acrylic—the mediator which is supposed to disappear from view and be unremarkable—that makes possible the seeing.

The technology that facilitates seeing animals has material and physical qualities that affect the way the animal is viewed. Putting this technology in view, especially when it is designed to be invisible, is necessary to analyze both the gaze between human and animal and how technology mediates the human-animal encounter. Technology is critical to the human-animal relationship, enabling close contact. The visitor in the aquarium (whether within the aquarium walls or online) can come to know and connect with the fish in the water only because of technology. Glass (and later acrylic) enables, situates, and magnifies human-animal relations in the aquarium, allowing them to a share space that would otherwise be impossible.

CHAPTER 10

TOMORROW'S MORE-THAN-HUMAN MINERS OF THE DEEP

CHARITY EDWARDS AND AMELIA HINE

Contemporary debates regarding this increasingly urbanized world bring attention to overlooked processes that extend beyond what is conventionally assumed to be the physical container of "the city." In considering the far-reaching influence of cities around the globe, the mining industry is a particularly pertinent case study for how such processes are supplied by interventions in ever more remote and challenging spaces, including the deep sea. In this chapter we argue that distant manifestations of urban life are made explicit through human-animal relationships transformed through technology.

Facilitating the expansion of these logics into the benthic realm is the "instrumentation" of those who are other than human. Indeed, the imbrication of humans, animals, others, and technologies in extractive practices required to sustain those conditions are the very futures that Donna Haraway might remind us desperately need "troubling." By way of example, southern elephant seals are increasingly utilized as bio-logged operators in underwater environments through technological augmentation (most often, gluing surface-mounted sensor and satellite relay instruments to their bodies), which enable humans to remotely conduct underwater research and more accurately profile the global ocean. Within challenging polar conditions, southern elephant seals are adept at the deep dives required to investigate these difficult terrains, which are still largely unexplored by humans. Conveniently their large bodies are also easy targets for scientific researchers to tag back on land with devices such as the Conductivity-Temperature-Depth Satellite Relay Data Logger (CTD-SRDL). At their most perfunctory the southern elephant seals represent a highly skilled and evolutionarily adapted "mobile infrastructure" at home in remote ocean environments that would typically call for more expensive and unreliable human-made hardware to undertake similar exploration tasks.

While the instrumentation of southern elephant seals for data gathering activities has been advancing steadily, humans have also been improving the technological capabilities of *machinic* exploration colleagues: specifically, autonomous underwater vehicles (AUVs). As seabed mining becomes an increasingly attractive option in our contemporary resource-restrained contexts, we contend that the primarily scientific research drivers of oceanic exploration will inevitably pivot to rare mineral extraction from the deep sea. Indeed, as law of the sea scholar Pradeep Singh notes, companies have been developing extractive prototypes for several decades in *anticipation of* actual mining.[1] Despite the southern elephant seal's hydrographic data collection remaining quite separate from the deep-sea mining industry today, its spatial and functional convergence with AUVs seems likely in the not-too-distant future.

Through this chapter we highlight nonhuman agency within a deliberately hybridized framework of remote oceanic exploration and an ocean entangled with urban processes operating at a planetary scale. With these relations in mind, we work toward recognizing and reframing the southern elephant seal and the AUV *beyond instrumentation*. As such we take a mixed methodological approach to our speculative inquiry: combining mapping and imaging practices to explore marine mammal–technological overlaps and performativity. Building on these inquiries, we employ the graphic and storytelling techniques of speculative and science fiction to project forward an imagined future of deep-sea mining that focuses on experiences of the elephant seal, the underwater robot, and their interrelationships.

These techniques highlight the ocean as a dynamic urban space and global entity. Selected images from current reportage and online sources readily capture existing phenomena, while *critical image-making* undertaken via collage and speculative cartography reveal neglected connections between the deep sea, urban processes, and the ever-changing dynamics experienced by various bodies within them. Image creation techniques like this demonstrate the conceptual work needed to break down prevailing boundaries that separate land, ocean, city, and planet. For this chapter critical image-making is a form of heuristic inquiry that has been developed in step with written theorizing to actively *re*-present the ocean. Human geographer Lauren Rickards writes of the importance of theoretical metaphor when reconceiving the world, but also warns these provocations may disappear once normalized, ossifying novel knowledge as literal fact.[2] Therefore, as Haraway also demands in this age of disruption and uncertain futures, we need continual new *imaginings* where other practices, histories, systems, and interactions are exposed and invite closer examination.[3]

Within these new imaginings, future more-than-human operators already seem destined to serve as yet another example of local knowledges appropriated by transnational mining companies to secure the resource speculation that far-off human populations depend on: a deep-sea environment characterized as "new ground," with oceanic spaces and benthic communities reconfigured for the de-territorialized state reliant on extractive capitalism.[4] To counter this inevitability, we ask how new relationships forming among humans, animals, and others may also *transform* spaces around the globe and, as urban geographer Kate Derickson prompts us, to "identify and nurture new solidarities and subjectivities, while troubling existing representations."[5] Drawing from the writings of Georg Simmel and Henri Lefebvre through contemporary scholars such as Doreen Massey, Judith Butler, Andy Merrifield, Ash Amin, and others working across geography, political philosophy, and planning disciplines, we note that these solidarities and subjectivities hinge on an understanding of the term "urban" that does not simply designate a geographic density of humans, buildings, and infrastructure but, rather, is conceptualized as the socio-spatial condition of sharing space with strangers. Moreover, reframing relations in the ocean mediated through technology can illustrate unexpected manifestations of urban environments as *strangely-shared spaces*. These include the revelation of once-presumed "wild" environments co-opted by human control *but without* human presence, and consideration of animals and machines as colleagues and cooperators in emerging labor logics that operate planetarily and increasingly structure life in often disregarded geographies. In this endeavor, we take a broad definition of "stranger" as one who exists in the "house" of another as either guest, visitor, or intruder—and do not presume to limit the stranger to humans or the house to just buildings.

DEEP-SEA URBANIZATION

We situate the problematic of our chapter in an ocean that has long been urbanizing, also making use of recent planetary urbanization debates in urban studies and human geography.[6] In keeping with those arguments originally developed by Neil Brenner and Christian Schmid (who themselves acknowledge a significant conceptual debt to Lefebvre), we also reject the reductive definition of contemporary life as binary conditions of either urban–rural or city–wilderness.[7] Urbanization should be recognized as planet-scaled: operating beyond conventional physical limits and manifest throughout seemingly remote environments, not all of which are dominated by *human* bodies.[8] Avoiding this understanding only continues to obscure destructive practices that escape city boundaries and work to transform landscapes that stretch across the planet with little oversight or interrogation.[9]

Significantly, even though these arguments have enabled some scholars to enlarge the conceptual frame of urbanization, its adherents have only more recently begun to explore the implications of such thinking for spaces like the ocean.[10] For instance, geographer Philip Steinberg and legal theorist Henry

Jones both remind us that oceanic space is far more likely to be seen as a kind of flattened void, delineated only by abstract geometries of imperialism, sovereignty, and capitalist exchange.[11] The ocean is much more than just an infrastructural surface; rather, it is a lived space of embodied, experienced, and materially unique territories encountered by humans, animals, and others in overlapping, uneven, and unexpected ways.

Consideration of an ocean that exists in relation to both far-reaching urban processes and the intimate mediation of technologies and human, nonhuman, and other bodies is also complicated by the physical obstacle that oceanspace presents more generally. It is enormous in scale and distribution across the planet, barely explored (by humans), always moving, governed at a distance via overlapping legal regimes, consisting of a materiality resistant to surveillance, and constituting a space generally unsuited to human life. Importantly, Lefebvre urges us to interrogate spaces that limit how we can conceive of urban life, regarding those parameters as physical and ideological "blind fields."[12] The construction of places and practices such as these seek to mask the formation of socio-spatial relations and therefore, he argues, we should dismantle prior frameworks, piece together unexpected and fragmentary information, and reveal the difference inherent to them.[13] To Lefebvre, this work requires "gathering together of what gives itself as dispersed, dissociated, separated, and this in the form of simultaneity and encounters."[14] By adopting this approach, we argue it is possible to explore how certain urban processes extend into the ocean, and where they are mediated through technologically driven encounters between humans, animals, and others.

Spaces such as the benthic realm where southern elephant seals and AUVs cohabitate are constructed on presuppositions of what we can and cannot see; so too, *remote* territories are produced by marginalizing or masking others.[15] Whether living or not, others are always subject to interplays of power and representation. Mining speculation and scientific research occurring in remote locations are not isolated operations but instead develop in the interests of human populations and their cities much farther away. How spatial relationships in the ocean and the seabed come to be represented are connected with the legibility of objects, practices, and processes in and of the world.[16] More specifically, urbanization is both supported and extended by resource extraction, and new mining techniques are expanding into even the most remote and challenging spaces of the deep sea. Such interventions are typically excised from considerations of more-than-human relations in the ocean, but we argue they should be revealed as emergent urban transformations occurring at the scale of the planet.

UNDERWATER EXTRACTIVE DEVELOPMENTS

Contemporary life comes with an accelerated demand for raw materials that enable technological innovation and the expanding influence of cities to continue. Alongside this growth comes the conflicting knowledge that resources contained within our planet—this Spaceship Earth—are finite and becoming ever more difficult and financially untenable to extract.[17] As the more accessible ore deposits close to the Earth's surface become depleted in heavily mined areas, the extractive industry moves toward high-quality resources much deeper in the subsurface and spreads into underexplored or remote locations in order to find and exploit new deposits of shallow ores.[18]

Deep-sea mining is premised on the availability of certain minerals and metals found on the seabed, like polymetallic nodules, hydrothermal sulfides, and ferro-manganese crusts.[19] Although discovered by scientists in the late nineteenth century, polymetallic nodules in particular have drawn more serious attention since technological developments in the 1960s made their extraction a distinct commercial possibility, and as their variable compositions of manganese, copper, nickel, cobalt, molybdenum, titanium, lithium, and rare earth elements are notably more lucrative today.[20]

Today the presence of rare earth elements is of keen interest to highly urbanized nations like the United States, with the majority of accessible terrestrial deposits of them located rather inconveniently within China's national borders. The prospect of guaranteed access to a secondary supply of these elements is thus highly desirable, especially as they play a crucial role in high-tech equipment manufacturing processes. They are the most valuable ingredient for products like smartphones, sophisticated military-grade vehicles, and renewable energy infrastructure, all of which structure contemporary life across the globe. Indeed, many populations are disadvantaged by lacking access to particular minerals as a result of geological "chance" dictating their distribution across the planet. India, for example, does not have any nickel or cobalt deposits within its boundaries.[21] The ocean thus offers a potential new source of mineral reserves and, by implication, the resource independence required for a range of geopolitical, economic, and other development ambitions.

ADVANCING TECHNOLOGY, ADVANCING AUTONOMY?

Underwater vehicles have been in development since the mid-twentieth century, in seeming sync with a growing awareness of the commercial possibilities of polymetallic nodules. Their origins lie in certain industrial applications derived from defensive ocean exploration and undersea rescue-and-recovery operations,[22] and several nations have spent significant funds over the decades advancing the military capabilities of underwater vehicles in particular.[23] Initially designed as remote underwater vehicles (ROVs), these objects were controlled by human operators located onshore or a nearby surface vessel and, significantly, remained tethered. Further research resulted in the creation of AUVs, freed from both physical tether and full oversight by human operators.[24]

Autonomous vehicle design has become a keen area of research for mining companies more recently. The Mine of the Future™ strategy announced by the Anglo-Australian mining conglomerate Rio Tinto Group encapsulates a desire to increase efficiency, safety, and production through total automation of mining sites, and has been enacted in current iron ore operations through the remote Pilbara region of northern Australia via a fleet of more than sixty automated mega-haulage trucks.[25] Although the extractive industry continues to invest large sums in broad research around autonomy, the application of such technologies are most pressing in remote areas where human access is limited or precluded entirely. Geographically extreme zones like Antarctica, the moon, outer space, and the deep sea are challenging for the human capacity to live, work, and potentially profit, in both physical terms and the current state of international legal instruments. However, with the recent resurgence of interest in deep-sea mining, these two areas come into more frequent collision and are driving research into improved AUV performance and reach. It is fascinating to recognize these ambitions articulated through sustained AUV research programs that openly proclaim their combined commercial and scientific research interests.

Much focus has been trained so far on the techno-capital potential of AUVs but urgent matters for future resolution include improving data collection, detailed oceanographic analysis, and accurate documentation of hydrographic profiles. A limited range of knowledge about the planet's seabed conditions—and especially in remote polar environments—continue to present challenges for both scientific communities and commercial industries. Humans have become increasingly reliant on AUV capabilities for remote research, even though ongoing obstacles are created through their sometimes erratic performance. In fact, a recent Weddell Sea expedition was called off when their high-value AUV "escaped" the digital oversight of its wranglers aboard the oceanographic research vessel, *S.A. Agulhas II*. The expedition had been searching for Sir Ernest Shackleton's lost ship, *Endurance*, which had been crushed by ice and sunk during an unsuccessful attempt to cross the Antarctic continent during 1914–1916. In 2019 extreme weather during the last days of the summer season trapped the AUV under a thick sheet of Weddell Sea ice, and the contemporary expedition feared meeting a similar fate to Shackleton's ship if they stayed to relocate the vehicle that had slipped its virtual lead.

The expedition director decried the challenging undersea environment as "the worst portion of the worst sea in the world,"[26] as Shackleton had before him. The AUV had been collecting high-definition color footage of the seabed—precisely where the *Endurance* itself may be resting still—and had also completed the world's (then) longest under-ice survey of more than thirty hours. These secrets are now lost, along with the vehicle, although several weeks of critical data had been transmitted back to base before the fateful final journey.[27] The unexpected loss of the *S.A. Agulhas II*'s underwater research vehicle is not

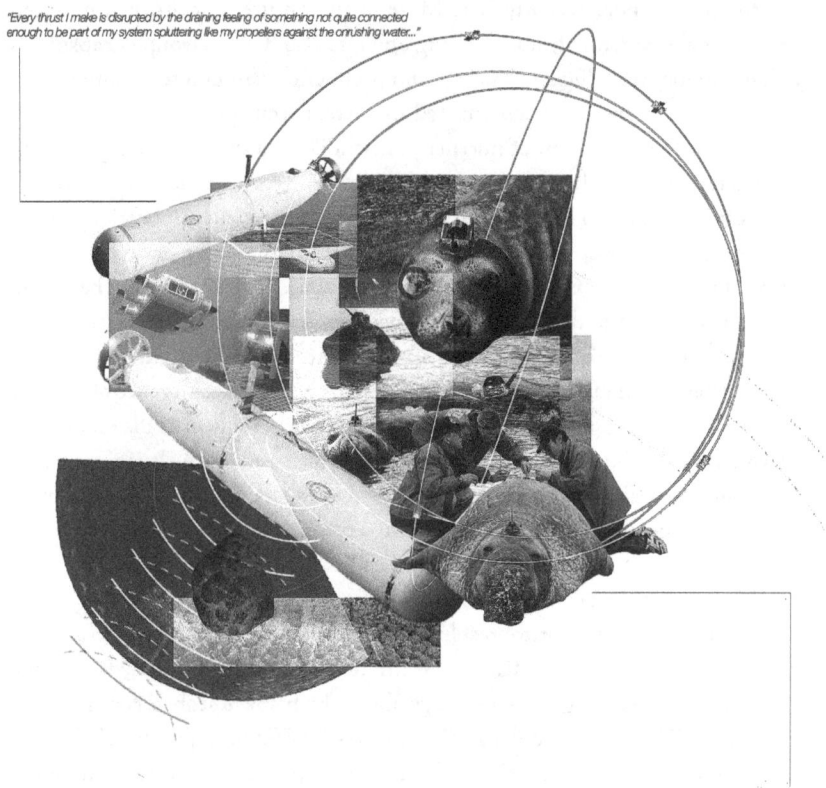

"Every thrust I make is disrupted by the draining feeling of something not quite connected enough to be part of my system spluttering like my propellers against the onrushing water..."

"Every dive I make is disrupted by the dragging feeling of something not quite smooth enough to be part of my form vibrating like my whiskers against the onrushing water...."

FIGURE 10.1. Antarctic operators of the deep sea. Illustration by Charity Edwards and Amelia Hine.

the only example of such objects escaping their supervisors. AUVs have been recorded as lost due to strong currents sweeping them off course from their human handlers' digital purview and also as a result of collisions with slow moving icebreakers.[28] Clearly human exploration of such complex zones is treacherous but even relatively sophisticated AUVs cannot yet match all the demands of extreme depth, distance, and temperature. Instead scientists more often achieve similar risk-laden research goals in remote environments by co-opting other living hosts (such as the southern elephant seal) through technological augmentation to engineer a more reliable category of autonomous colleagues for these tasks.

MORE-THAN-HUMAN OPERATORS OF DEEP-SEA URBANIZATION

Southern elephant seals are utilized as augmented operators in deep-sea environments, allowing humans to remotely conduct oceanographic research at a distance through the deep dives required to investigate difficult underwater terrain and the subsequent acquisition of "pole to pole" hydrographic profiles. On dry land southern elephant seals are far less nimble and make especially easy targets for augmentation by scientific researchers. Once tagged with satellite-linked bio-logging devices, they are enrolled underwater as adept workers—far more so than humans or even AUVs—in the service of exploring deep seabed environments. Difficult locations usually call for expensive and delicately calibrated hardware to undertake such tasks, but these seals function as an evolutionarily adapted "mobile infrastructure" of sorts in the hands of their human supervisors and at a mere fraction of the costs ordinarily associated with such labor. Southern elephant seals were the first marine mammals to be employed as "oceanographic samplers," fitted out with the contemporary technology of CTD-SRDLs. The species was chosen primarily because of its ability to dive continuously to great depths for significant periods of time[29] and its chosen habitat, which encompasses extremely remote Southern Ocean areas south of the Antarctic Circumpolar Current (as illustrated in fig. 10.1). The CTD-SRDLs attached to southern elephant seals are designed to measure vertical profiles of temperature and salinity for a better understanding of oceanographic processes, and they also contribute new knowledge of the seals themselves in terms of foraging, reproduction, and population changes.

Fitting the loggers to individual southern elephant seals and the ethics of their subsequent "instrumentation" is a practice not widely discussed within the scientific literature, which admittedly focuses more on the results of hydrographic mapping. A recent study into the habitat information acquired from such data, however, does blithely detail the process of capture, fitting, and deployment: "Two hundred and eighty-seven southern elephant seals were captured at eight deployment sites around the Southern Ocean each year between 2004 and 2010 . . . at the end of their annual breeding haul-out. . . . The seals were chemically sedated . . . , weighed, and measured . . . , and a CTD-SRDL-9000 . . . was attached to the hair on the seal's head. . . . We are confident that the instruments did not affect their at-sea behavior given that the smallest instrumented seal weighed 169kg, making the tag <0.3% of the seal's weight."[30] While we naturally question weight change as a defining factor for behavioral impacts from bio-loggers, our overriding concern is with the transformation of marine mammals from agential beings to augmented technologies. The southern elephant seal effectively becomes a "more-than-human operator" during this process, lending its capabilities to human-led research with an uncertain end point, as explored above in relation to other uneasy commercial-research partnerships. Indeed, as marine biologist and animal-borne technology researchers Boehme and colleagues also recently noted, the perfunctorily named Southern Elephant Seals as Oceanographic Samplers project deployed these augmented marine mammals in the Southern Ocean to gather well over twenty thousand oceanographic profiles across several years, an extraordinary amount of scientific research labor that "represented a major addition to the World Ocean Database."[31] Although animal use in scientific and product testing laboratories has declined significantly in the last few decades, emerging biotechnology and animal augmentation methods signal that "animal economies" are fundamental to extractive capitalism across the planet.[32] These actions also highlight compelling legal, ethical, and conceptual quandaries that arise from new life-forms[33] and, we argue, especially those produced from the intersection of scientific research, technological innovation, and commercial imperatives such as industrialized seabed mining.

Augmented southern elephant seals are not simply describing bathymetry or recording observations as per the physical scientific methodologies mentioned above. They are *performing* those terrain explorations simultaneously during their everyday activities and often correcting the anthropocentric hydrographic record. These new "bio-technical arrangements for imaging and observation"[34] operate in line with human goals for exploration and experimentation, but they do not necessarily limit or exclude nonhuman agency. The appropriation of the "citizen scientist" labor undertaken by southern elephant seals does signal how rarely we consider other animals as *colleagues*, collaborating across the (supposed) species divide in practices of knowledge production. This is mirrored by autonomous undersea vehicles, which extend human intent and endeavor into remote environments but without the actual presence of humans. Admittedly, these *machinic* knowledge workers have a different kind of agency from the southern elephant seals. With a capacity to draw on project parameters and dynamically respond to unique spatial conditions to achieve identified outcomes, AUVs are perhaps more embedded within decision-making structures (and thus power relations) of mineral resource exploration. In either instance however, it is clear that a range of animals and objects are optioned where human science and commercial hierarchies see fit for the purpose at hand.

Conventional conceptions of what is and what is not considered "urban" serve to mask practices in remote areas that both stem from and reproduce human-nonhuman estrangement alongside deleterious environmental impacts extending beyond the physical footprint of any one city. More to the point, geographer Chris Philo argues that securing the nonhuman through science and technology as an available resource for human domination—in this instance, the seabed and mineral bounty located within it—inevitably leads to fundamentally *insecure* environments at much larger scales.[35] Obscuring this within the deep sea runs counter to an urgent need to respond to destabilizing transformations experienced in contemporary life. This is typically recognized in rampant habitat loss, polluted water systems, and ever-growing extractive industries around the world, but rarely does our perspective consider animals within the labor logics that increasingly structure planetary relations from pole to pole. Instead, popular perception of the deep sea (including the Southern Ocean diving habitat of southern elephant seals) tends to reinforce constructed hierarchies found back on dry land, which themselves sustain "hegemonic human/nonhuman binaries."[36] Aligning our investigation in this chapter with Jane Bennett's reference to a "confederation of things,"[37] we seek to counter these unhelpful distinctions. We maintain that emerging assemblages of humans, animals, objects, techno-scientific endeavors, and largely disregarded ocean environments are due much closer consideration. In doing so the more-than-human possibilities of *deep-sea solidarity* might allow a radical reconsideration of processes currently transforming our planet.

SPECULATIONS ON THE MINES OF TOMORROW

In highlighting such possibilities and engaging with their capacity to reframe urban processes infiltrating previously remote territories, we draw from Anna Tsing's extended Strathernian method of "reification for the work of comparison."[38] Tsing's method involves treating nonhumans as ethnographic subjects and engaging them as devices that offer a "critically reflexive view of our tools for knowing action and agency."[39] Through these acts of attentive regard—looking and thinking closely with nonhumans—we are able to provide "'surprising comparisons' or 'cultural analogies' crafted to take us up short, forcing us to rethink how we think"[40] and enable reflection on the limits of our own ways of seeing the world. The following vignettes aim to do this work by taking us underwater with our instrumentalized nonhuman and machinic research colleagues, facilitating moments that draw attention to how life and interactions might otherwise play out. In doing so these vignettes, told from the perspective of the elephant seal, make the ocean visible as a site of complex power relations.

> *Taking a breath and aiming my hydrodynamic body down toward the seafloor, I am relieved to be able to start my dive back into the dark depths of the ocean and away from the never-ending glare of the Antarctic sun. Although I am eager to be on my way down into the saltier, denser waters in search of food, I can feel the same irritating tension at the base of my skull that has been on my mind for the last few days. Every dive I make is disrupted by the dragging feeling of something not quite smooth enough to be part of my form vibrating like my whiskers against the onrushing water. Although it's hard to know where it came from, it's a new feeling, and one that has taken time to adjust to. Despite this distraction, my body performs its usual transformation from land to sea-dwelling mammal without issue.*

> *As I swim deeper through the depths of the mesopelagic zone I feel the increasing pressure of the water around me and my lungs begin to collapse. The oxygen I have stored in my blood and muscles will keep me alive for a long time, however, and my heart rate has slowed by three quarters while I hold my breath, preserving the*

oxygen further. My lungs will reinflate when I get close to the surface. This is a clever mechanism that my kin have perfected over a great many generations. We dive deeper and longer than most mammals, except for some whales, and it's always a shock to come across one of those in the deep darkness.

Moving down into the bathypelagic zone, my eyes adjust quickly to the absence of light, and I begin to search for squid. It's been getting harder to find the particular squid that I like, and often I end up eating only fish. My kin and I worry the lumbering creatures we keep meeting on the seafloor have scared the squid away. We have watched these brutes moving slowly closer to our subantarctic rookeries as they scrape up the top layer of mud and life from the seabed, throwing sediment around and digesting all the microorganisms living on the metallic lumps they collect.

Moving ahead of them, like curious scouts, their yellow cousins light up the dark and blind us when we least expect them. My eyes are sensitive to the short wavelengths of the bioluminescent animals that float in this depth, letting me forage using a combination of their light and my vibrissae. When the yellow fish appear suddenly I often lose many minutes of my dive time waiting for my eyes to readjust to the bioluminescence. While they are much larger than me, they never seem to present a threat; they carefully navigate around my body whenever we meet and continue busily on their path until they fade back into the black and my eyes can readjust. I have heard my kin tell stories of other types of yellow fish that are more dangerous than these bright and careful ones, leaving whales and other creatures disoriented and damaging their hearing.

As seabed exploration facilitates increasingly feasible options for rare mineral extraction, the convergence of southern elephant seal and AUV functionality seems inevitable, perhaps even replicating existing mining venture practices that co-opt local communities and traditional knowledges in order to aid speculation and resource location. These processes are carried out in the interests of growing human populations around the world even where they operate in the absence of human bodies in remote environments. We ask: Can we reframe the southern elephant seal and autonomous underwater vehicle beyond mere instrumentation and speak to the agency of their future working roles in these endeavors? And in doing so might we be able to recognize these more-than-human operators of the deep seabed as unexpected collaborators? That is, what relationships form between humans, animals, and others at the seabed, and, as per Derickson's earlier advice, what new solidarities and subjectivities can we support by acknowledging these possibilities?[41]

Emerging relations between extractive processes, humans, and others are fundamentally transforming spaces at a planetary scale and can be seen clearly enough through resource speculation within industrialized modes of fishing and land-based agriculture. Seabed mining also has the capacity to both support and extend extractive capitalist economies that themselves sustain cities across the planet. We argue, however, that by troubling existing representations of southern elephant seals and AUVs of the deep seabed, we might forge a more optimistic reading of an ocean entangled with urban processes, especially where human presence is so readily compromised by our own physical inability to dominate this space.[42]

We call on Haraway again as we engage ever more deeply in demarcating the Capitalocene, particularly where she advises, "One must surely tell of the networks of sugar, precious metals, plantations, indigenous genocides, and slavery, with their labor innovations and relocations and recompositions of critters and things sweeping up both human and nonhuman workers of all kinds."[43] In this inquiry, we suggest that networks of surveillance, speculation, and data acquisition produced by more-than-human colleagues of the deep sea are due the same consideration and, importantly, *speculation* into futures that develop in response to these troubled representations of the southern elephant seal and AUVs in confederation. Within such scenarios, our own human eye functions are supplanted by image practices that no longer refer to a "real" observer located in our perceived world.[44] Paul Virilio in *The Vision Machine*[45] and many others reference examples that articulate where vision is unmoored from human observation. Indeed, existing representations of the seabed appear to limit our conception of bodies and things at work in the oceanic volume and only reproduce physical and ideological enclosures of space at the scale of the planet. Let us therefore recalibrate that vignette once more:

> *Snapping an image and aiming my hermetically sealed body back toward the sea ice opening, I am relieved to be able to start my ascent to the bright air at the ocean surface and away from the never-ending damp of the Southern Ocean deep. Although I am eager to be on my way back to the lighter, clearer surface in search of recharging, I can feel the same irritating tension at the base of my communication circuits that has been in my mind for the last few hours. Every thrust I make is disrupted by the draining feeling of something not quite connected enough to be part of my system spluttering like my propellers against the onrushing water. Although it's hard to know where it came from, it's a new feeling, and one that has taken time to adjust to. Despite this distraction, my body performs its usual transformation from undersea to under-ice vehicle without issue.*

> *As I swim higher beyond the depths of the bathypelagic zone, I feel the decreasing pressure of the water around me and my thrusters begin to relax. The charge I have stored in my fuel cells and circuits will keep me active for a long time, however, and my photosensor collection rate has slowed by three quarters while I hold my course,*

preserving the energy further. My thrusters will reorient when I get close to the surface. This is a clever mechanism that my kin have engineered over a great many iterations. We dive faster and longer than most vehicles, except for some submarines, and it's always a shock to come across one of those in the deep darkness.

Moving up into the mesopelagic zone, my sensors adjust quickly to the presence of light, and I begin to search for a sea ice opening. It's been getting harder to find the particular opening that I need, and often I end up swimming only farther. My kin and I always worry that the lumbering creatures we keep meeting at the sea surface have destroyed the ice. We have watched these brutes moving slowly closer to our deep-sea launches as they scrape the layers of ice and life from the ocean, projecting acoustic signals around and alarming all the organisms living in the deep they observe.

Moving around them, like curious scouts, their grey cousins speed up in the dark and bypass us when we least expect them. My detectors are sensitive to the steady wavelengths of large objects that float in this depth, letting me avoid collisions using a combination of their predictability and my propellers. But when the grey bulks appear, I suddenly, I often lose many minutes of my dive time waiting for my detectors to readjust to their presence. While they are much faster than me, they never seem to present a threat; they carefully navigate around my machinic body whenever we meet and continue busily on their path until they fade back into the black and my detectors can readjust. I have heard my kin tell stories of other types of moving bulk that are more dangerous than these small and careful ones, leaving submarines and other technologies disoriented and damaging their communications.

In producing a deliberate comparison between southern elephant seal and AUV experiences of oceanspace through these two sets of vignettes, we highlight similarities in terms of their spatial occupation and roles in surveillance, but also draw out important differences. In particular, the vignettes reveal the southern elephant seal as a *living* being that is easier to imagine as a self-contained agential organism, more so than the algorithmically determined oceanic exploration of the AUV at least. In positioning the AUV's movements in relation to, and partial mimicry of, the southern elephant seal, we offer an opportunity to "rethink how we think."[46] How should we think about the agency of machines such as AUVs if comparing them to *living* beings runs the risk of further conceptually instrumentalizing animals? We contend it is possible to think about the AUV in terms of vital materialism: agential in the sense that its material components have come together from various corners of the globe via manufacturing processes to assemble briefly as a single, self-contained robot. Each of those components is created by different

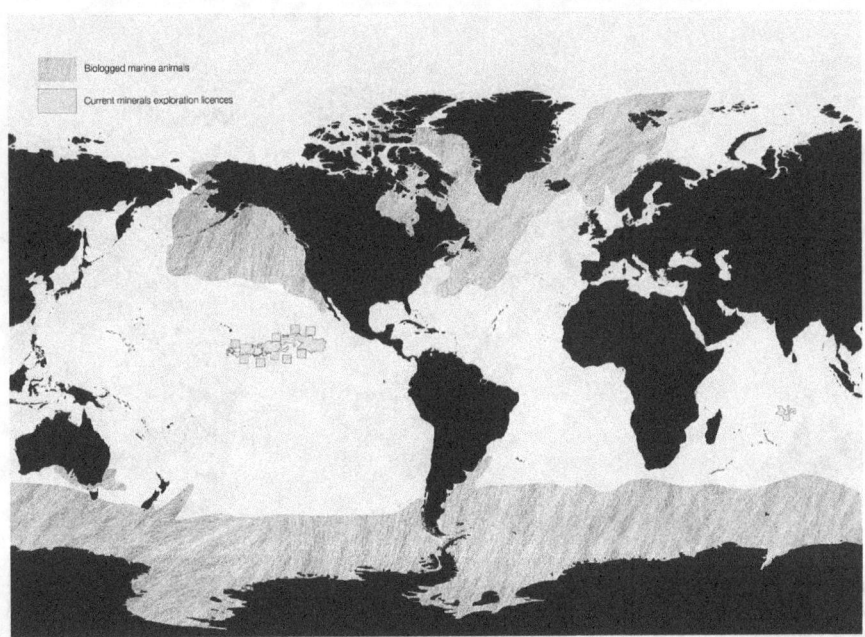

FIGURE 10.2. Existing horizontal geographies: current location of mineral exploration licenses, and the geographic extent of logged marine animal exploration. Illustration by Amelia Hine.

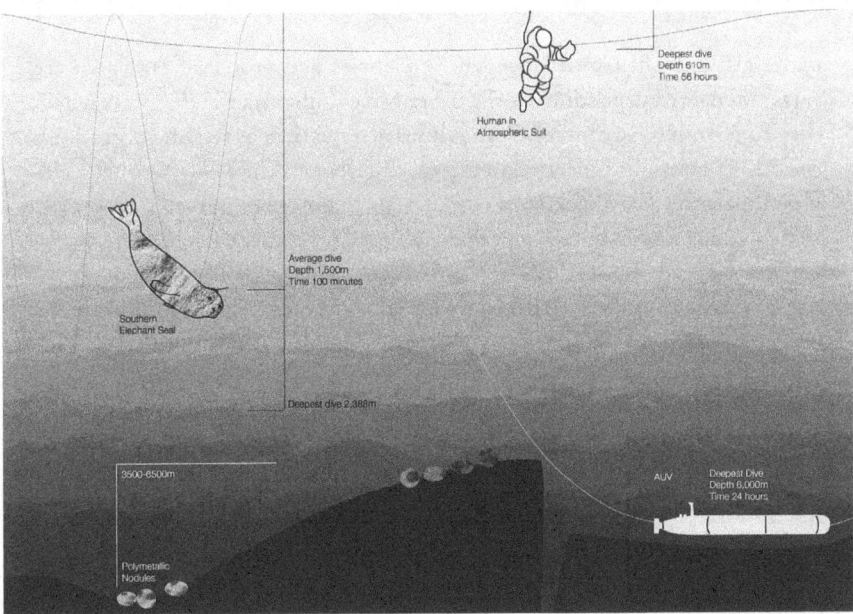

FIGURE 10.3. Existing vertical geographies: depth and time convergences in vertically remote ocean space exploration. Illustration by Amelia Hine.

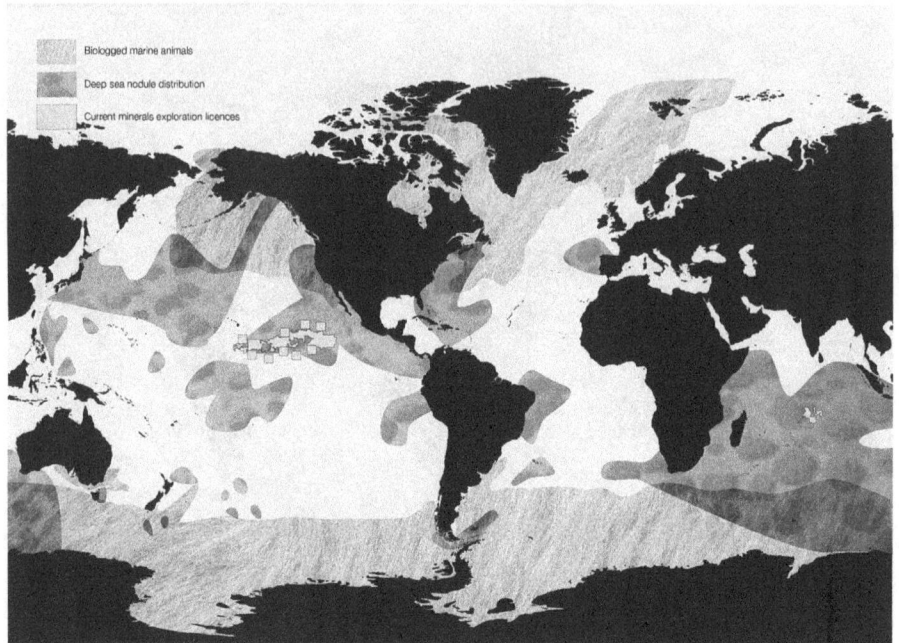

Biologged marine animals

Deep sea nodule distribution

Current minerals exploration licences

FIGURE 10.4. Speculative horizontal geographies: the future extension of deep-sea mining into less concentrated areas of polymetallic nodule distributions across the seabed. Illustration by Amelia Hine.

materials with their own "incipient tendencies and propensities, which are variably enacted depending on the other forces, affects, or bodies with which they come into close contact."[47] This approach requires us to radically rethink the AUV: instead of inert and entirely different to the *living* southern elephant seal, they are constructed of materials that have converged temporarily as nonhuman bodies or self-contained "objects," "because their becoming proceeds at a speed or a level below the threshold of human discernment."[48] We do not intend to suggest that such material convergences are not individuals, but rather advocate for shifting conceptions of the AUV from inert and inanimate to a complex and dynamic nonhuman typology whose instrumentalization we should question in a similar way to organic life-forms such as the southern elephant seal. At the same time, however, our ability to get into the head of the seal is in a sense just as limited as that of AUVs. While we share many traits—warm bodies, sensory knowledge, sentience, and consciousness, for example—it is impossible to really know the mind and therefore the lived experience of the elephant seal. As such, our use of Tsing's Strathernian method applies just as readily to the seal as to the AUV.

The following cartographic overlays (figs. 10.2, 10.3, and 10.4) all register a speculative convergence of such beings from their current global distribution in potential futures, should deep-sea mining develop further and the International Seabed Authority relax its current regulatory grip. These intersections occur geographically and manifest in terms of shared capacity to withstand great depths and high pressure for significant periods of time. Deep-sea solidarities carve out a shared vertical geography that, while not necessarily remote from a global perspective, is an extreme environment beyond usual human tolerances. The initial mapping series (shown in fig. 10.2) concentrates on traditional horizontal geographies, identifying the current location of minerals exploration licenses and bio-logged marine animals. Figure 10.3 accompanies this map of existing geographies with a vertical section of the ocean indicating depths and temporalities of humans and nonhumans underwater, giving an indication of the points at which physical intersections may be possible beyond the horizontality of the map.

Although it is not possible to gather data on the locations of all current commercial AUVs, it is a safe assumption that current mineral exploration sites are utilizing a combination of human operated vehicles (HOVs), ROVs, and AUVs. Similarly, the data captured by Marine Mammals Exploring the Ocean Pole to Pole (MEOP)—an international database of mapping undertaken by CTD-SRDL logged sea animals—are not exclusively from southern elephant seals, although the majority of the Southern Ocean data can be assumed to be. This map (see fig. 10.4) therefore expands the exploration zone into a speculative future, with the onset of deep-sea mining, improvements in technology, and the rapid depletion of minerals in the current exploration areas forcing expansion into areas with less than the required 15 kg/m² wet weight for economic viability.[49] Declining ore grades, new exploration areas and technologies, and "the development of large scale-low cost mining fleets" are some of the factors that have previously led to the expansion of extraction zones, both vertically and horizontally, within traditional terrestrial mining industries such as gold mining.[50] It is no big stretch to imagine similar factors influencing deep-sea mining expansion.

An ocean entangled with processes that operate through human control but without human presence reminds us to stay conscious of both the intermediaries who facilitate vision and unlikely colleagues who remain obscured from view. Beyond mere slippages of surveillance, representation, and control driven by Google Earth and associated satellite imagery, we should recognize the unusual nature of data gathered by machines and others such as the southern elephant seal. As has often been the case with other forerunner imaging technologies, there is a danger these new visioning beings will have their data conflated with *actual human vision*. As such, they stand to lose not only recognition of their individual agency—in service of the "View from Nowhere," or (more disparagingly) the "God Trick"—but their very "thingness."[51] Ultimately, by ceding the autonomy of data-gathering and image-making to other

FIGURE 10.5. The deep-sea urban, unmoored from human participation. Illustration by Charity Edwards and Amelia Hine.

life-forms, we might better understand the often unexpected implications of processes largely unmoored from human participation, as is demonstrated in the speculative vision of figure 10.5.

An intriguing additional complication at this stage concerns bio-hacking risks for instrumented animals and autonomous vehicles in the deep sea and the possibilities for novel forms of seabed piracy in response to mining ventures that rely on the support logistics of other life-forms.[52] This fear is further escalated when considering the likelihood of resource conflicts at such sites and the potential for illegal abduction of augmented technologies or violence enacted against innocent colleagues at the seabed. Taken together, speculations on the mines of tomorrow and their more-than-human operators are a complex provocation. The few vignettes presented in this chapter all too clearly identify the failure of current debates to consider how technology radically alters the manifestation of urban processes in remote spaces like the deep sea. These speculations may therefore be key to understanding the entanglements of humans, animals, and, importantly, others with technology at a planetary scale.

By focusing on the agency of the southern elephant seal and the AUV and their potential for collaborative, and even collegial, relationships in extreme conditions, we hope that they can be reconsidered beyond mere instrumentation. There is great potential to expand a concept of agency beyond those animals and machines working within already recognizable labor geologics, and include less "human" more-than-humans in processes that manifest via technologies operating at the scale of the planet.

CHAPTER 11

GUIDE DOGS AND ELECTRONIC MOBILITY AIDS AS THE MEDIATORS OF ACCESSIBILITY

RIIN MAGNUS

Although technology helps to fulfill certain tasks, it is usually not considered in terms of assistance. "Assistive technologies" is a specialized term that designates "products, equipment, and systems that enhance learning, working, and daily living for persons with disabilities."[1] Similarly, the companionship of pets has been linked to diverse psychological and physical advantages.[2] However, these advantages are not commonly discussed in the context of assistance. The term "assistance animals" (or "service animals") refers to those animals that provide assistance to a person with a physical or psychological disability. While acknowledging the dependence of all humans on other species as well as technologies for making us who we are, this chapter sets its focus on assistive devices that are developed and service animals who are trained to assist people with vision impairments, exploring above all their role in the shaping of the *Umwelten*[3] of their users and handlers. Furthermore, the chapter will discuss how they complement or impede each other in this Umwelt formation process, consequently influencing the provision of accessibility.

Although electronic devices that are meant to complement guide dogs or white canes[4] are developed to provide their users with information that is not accessible otherwise, the fundamental differences in the functioning principles of different aids as well as in their presumptions about human cognition are often left without consideration. Prior research has predominantly focused on contrasting the social dimensions of diverse mobility aids[5] or evaluating the technical functionality and usability of such devices.[6] The following study aims to provide a semiotic account of the functioning of the two types of mobility aids, guide dogs and electronic mobility aids, discussing how they mediate environmental information, while also examining their integration into the fabric of social interdependencies.

Various assistive devices impose distinct requirements on both environmental cues and the cognitive capacities of their users. These demands span a spectrum from cross-species communication to acoustic spatial understanding. However, could the concurrent usage of multiple devices mitigate some of these requisites, or would it instead amplify the overall demands? What do the Umwelten of the users of different assistive devices look like? Furthermore, how might environmental designs incorporate affordances that facilitate user mobility within different environments, accommodating a wide array of mobility aids? All these questions ultimately converge on the focal point of accessibility. This concept encompasses more than just the availability of physical spaces; it also pertains to the ability to engage in and cultivate specific social relationships. Simultaneously, these inquiries address the challenges associated with the diverse array of species coexisting within urban landscapes. In the context of accessibility, this prompts an examination not only of how specific human bodies are prioritized through the design of urban environments, but also of the incorporation or exclusion of other species. This includes both animals that serve as human companions and those who have independently navigated their way into urban areas, affecting their access to particular resources. Understanding how such social affordances are entangled with physical affordances and how these, in turn, are coupled with the sensorimotoric and social capacities and interdependencies of the users of different mobility aids could lead to a more symbiotic development of different mobility aids, rather than prompting competition or neglect among them.

THE DEVELOPMENT OF GUIDE DOG AID AND ELECTRONIC ASSISTIVE DEVICES FOR THE VISUALLY IMPAIRED

White canes, the most popular modern mobility aids of the visually impaired, came into extensive use only in the 1950s.[7] However, nearly half a century earlier initial attempts were undertaken to develop mobility aids that incorporated signal transmission and transformation for those with visual impairments. One of the pioneering technical mobility aids, known as the Elektroftalm, was conceived by the Polish physician Kazimierz Noiszewski in 1897.[8]

The simple device consisted of a selenium photosensor placed on the head of the user and an acoustic emitter. The intensity of the sound was to correspond to the light conditions of the environment. Elektroftalm preceded later electronic travel aids which, though embracing a range of software and design approaches, operate on sonar-based, vision-based, infrared, or GPS sensors, as well as tactile or acoustic emitters.[9] Extensive development of electronic travel aids began after World War II together with the development of ultrasound, laser, and computer imaging technologies.[10] For a long time the devices meant to aid navigation, obstacle detection, or orientation of the visually impaired were developed independently,[11] resulting in complications for users who had to combine different devices for independent travel. Even these days different electronic mobility devices often serve different functions, and merging multiple functions into a single device has posed a major challenge for the developers of electronic aids. Given its functional specialization, a particular travel aid usually needs additional input from either other portable mobility aids, live companions (human or nonhuman), or the physical environment to guarantee independent movement. For example, GPS devices with acoustic output often need an additional obstacle detection device, although two-in-one devices with both GPS and obstacle detection have been developed.[12] Portable electronic devices are supported by the stationary location identifiers placed in the environment (such as talking signs or verbal landmarks emitting infrared signals when the user with the corresponding receiver is in the vicinity, or tactile maps helping in navigation, etc.).

In terms of their function, guide dogs are closest to obstacle detection devices that provide information about the proximate environment of movement. Yet although guide dogs are trained to independently notice and inform their handlers about obstacles and to follow the handler's instructions in finding the trail, the dogs, being living beings, also utilize their natural skills of orientation and navigation. Although most of the dog's decisions are generally approved by the handler, the differences between the dog's Umwelt and the handler's can sometimes lead to confusion. The limitations associated with using guide dogs as aids for mobility are sometimes likened to the limitations of white canes—in both cases only proximate obstacles are detectable, objects located above the head remain unnoticed and orientation information is unavailable.[13]

The professional training of guide dogs began a couple of decades after the introduction of the first electronic mobility aids, during World War I. The dogs were trained to help blind soldiers regain their self-esteem but, more pragmatically, to help them return to the job market. The recognition of guide dogs as mobility aids for war veterans differed from country to country. This disparity was influenced by prevailing perceptions of the dog's role as well as the specific attributes that visually impaired war veterans were expected to uphold in spite of their disabilities. For example, the introduction of guide dogs differed significantly in post–World War I Germany and Great Britain, as demonstrated by the study by Anderson and Pemberton.[14] In Germany, where the institutional training of the dogs began, guide dogs (at that time exclusively German shepherds) were included in various rehabilitation measures that the German state provided to war veterans, whereas in Britain the growth of the guide dog movement in the 1920s was relatively slow.

The acceptance of guide dogs as mobility aids for the war-blinded was slowed down in part by the central philosophies of the institutions focusing on rehabilitating them. As Anderson and Pemberton note, St Dunstan's Home, a charity that specifically supported those blinded in the war, upheld a "walking alone" philosophy as the ideal for visually impaired ex-soldiers, in which case "a dog denoted dependency and undid all the good work associated with them as cultural signifiers of heroism, triumph over adversity and the embodiment of a particular type of restored masculinity."[15] During the interwar period guide dogs were often presented as a kind of prosthesis, attached to the human and restoring their initial physical state.[16] However, the principles of training devised at that time demonstrate a tension between the recognition of the dog's individual and species-specific Umwelt and the dog's functioning as an attached prosthesis of the human being.

In the 1920s and 1930s Emanuel Sarris and Jakob von Uexküll pioneered a guide dog training known as the "phantom man" method.[17] A crucial aspect of this training involved altering the dog's perception of their body plan, achieved through the utilization of a compact cart fastened to the animal subsequent to initial obedience training. The dimensions of the cart, including its height and weight, were designed to simulate human proportions. As the dog maneuvered with the cart, she or he gained the ability to autonomously navigate around obstacles that held particular significance in the context of guiding visually impaired individuals, such as mailboxes and open windows. Following the cart phase, the training progressed to working with the trainer, who simulated a blind person, and culminated in training sessions alongside an actual blind person. Yet when moving with the handler, the dog was to continue with the set of signs and meanings she or he had acquired when moving with the cart, remaining relatively insensitive to the cues stemming from the handler's behavior. This approach was ultimately abandoned due to the dog's limited responsiveness to human cues. Although these days the methods and standards of training vary from school to school and country to country, the training of a guide dog always involves three major phases: socialization of the guide dog puppy with a "foster family," training at a guide dog school, and training with the designated handler.

Even though electronic aids and guide dogs have been available as vision assistance for over a century, their collaborative development and concurrent utilization have exhibited slow progress. When conducting interviews with guide dog users for my research about the sign usage of guide dog teams,[18] I found out that they are mostly reluctant to use any electronic mobility devices alongside their guide dogs, although some of them had tried to do so. According to my informants, the guide dogs successfully mastered all the necessary orientation and navigation tasks. If guide dog handlers did choose to use an additional aid, it was the traditional white cane or touch cane or occasionally also a GPS. The general critique of visually impaired persons targeted against electronic devices concerned the overload of information produced by these devices. A British guide dog user, who had prior experience with ultrasonic glasses, expressed her thoughts on their usage:

> I didn't find them very easy to use. I think for me they gave too much information about what was going on around me. . . . I have used other similar devices for testing purposes and with all those devices which gave you ultrasonic feedback and basically gave you detection from a laser, I just found that maybe with lots of people around there was just too much information to concentrate on my long cane or on my guide dog plus the noises and the physical feedback from the devices and to concentrate on where I was going at the same time. I found all that a bit too much.

A German guide dog user, who had used the Kay glasses developed in New Zealand, gave a similar opinion: "It's very tiring, very, very tiring, because you listen to the noises of the gadget and you listen at the same time to the noises of the street and the houses, the cars—everything you know. It's a salad of noises and it's very, very tiring. When you go with the dog, you concentrate on the things around you, to scan, so to speak, what is around you, and the dog solves your problems with the direction and the obstacles and so on. Very relaxing." Clearly social and psychological factors also come into play when deciding between different assistive devices.[19] Presumably guide dogs will continue to serve their function alongside the technical devices irrespective of the (info)technological progress that the electronic travel aids undergo. Moreover, the rise of a variety of novel service animal functions in the past decades, particularly in therapeutic and medical contexts, from seizure alert dogs to autism service dogs, suggests that animals are valued for their particular ways of making sense of the world and to communicate information to others. These animal-assisted interventions can, in turn, be integrated into complementary technological frameworks.

It is highly likely that the assortment of mobility aids and their combinations will endure into the future. Even if visually impaired individuals using different aids attain comparable levels of mobility and reach destinations with similar ease, they will remain situated within distinct Umwelten, resulting in varying requirements for environmental affordances.[20] These disparities in Umwelten stem from the distinct aspects of signal detection, information encoding, and interpretation that different types of devices facilitate for users. The necessity for diverse affordances consequently entails the need for specific environmental conditions and designs that align with the particular characteristics of each device. Nevertheless, the persistence of diversity and the resulting array of choices for vision aids should not hinder their development while taking this diversity into account. In this context information pathways are distributed, and the presence of hybrid sources for Umwelt formation is acknowledged. This has the potential to lead to a convergence of technology and organisms in shaping visual experiences, thereby enabling the development of design features that cater to such synthetic experiences.

SENSORIMOTORIC MEANING GENERATION

Electronic mobility devices are not intended to directly replicate the act of seeing, but rather to assist in performing specific functions that are typically associated with vision, especially within certain environments. The same principle applies to guide dogs: they are trained with a distinct focus on mobility-related functions, which then become more refined during the collaborative training and movement phase involving the user and the dog. Each form of assistance involves particular assumptions about the sensory and motor foundations of meaning creation, along with expectations regarding the specific abilities related to sign use.

Although at first glance it might appear that mobility aids operate by supplying supplementary sensory input to enhance the user's mobility, the interweaving of perceptual and motor processes while acquiring environmental information is substantially more intricate. Initially, mobility aids must anticipate the user's motor responses within the signals they convey. This implies that the sensory information transmitted by these devices takes into consideration both the user's physical attributes and their range of movement. The motor response is preconceived within the signal's structure. For example, some devices provide distance information first and subsequently transmit the information about the shape and texture of an object once the user approaches it. Among guide dog teams, the pace of movement itself mirrors the prevailing environmental conditions—slower in bustling and crowded locales, and swifter in open streets devoid of heavy traffic. A reduced pace also signifies a possible obstacle along the path or the proximity of the destination or object directed by the handler's instructions. While guide dogs are sometimes perceived to lack the ability to convey information about distant objects, this is frequently not the case. Guide dog owners can discern from alterations in the dog's pace and body tension whether an obstruction is nearby or if a significant and anticipated object lies in the path. As articulated by an informant from Estonia: "And when I issue a command while on the street, I can deduce from the dog's accelerated pace or if she begins to turn her head whether she is still in the process of searching. For instance, if I instruct her to 'Search for a zebra [crosswalk],' she starts looking for it, and upon spotting a zebra, her strides adopt a distinct tipping motion, as if to convey, 'I've located it, let's move swiftly now.'" Nonetheless, these signals generally remain quite broad, and supplementary electronic aids could potentially address the lack of specificity in movement speed.

Despite the correspondence between the dog's speed and the signal modulations in electronic devices, the behavioral response in the case of guide dog teams hinges on mutual coordination and joint decisions between the two subjects. Simultaneously, the movement of guide dog team members isn't solely dictated by environmental factors; it's rooted in a shared sensitivity to each other's mental and physical states. For instance, a guide dog user from Estonia, who is grappling with diabetes, shared that her dog's behavior differs significantly when her blood sugar levels are high, causing a reduction in pace. Another informant from Estonia recounted how weather conditions can impact the dog's condition and consequently the pace—allowing for swifter progress on cooler days, while necessitating extra encouragement on hot summer days.

The over fifteen thousand years of cohabitation between humans and dogs has resulted in what Donna Haraway has highlighted as "sympoiesis,"[21] and what Michele Merritt has referred to as "collaborative cognition."[22] Merritt has elaborated on her stance concerning the interdependent influence on the cognitive development of dogs and humans: "Much as the radicals in philosophy of cognitive science claim that some forms of cognition can be properly understood only by including the active environmental engagements as part of those cognitive processes, so too, I argue, do dogs and humans form collaborative pairs in which unique forms of cognitive processes emerge."[23] Moreover, sensitivity to human cues and the ability to use those cues to guide further actions (sometimes referred to as the humanlike social skills of dogs)[24] stand out as prime rationales for the unparalleled efficacy of dogs in the guiding role. Dogs are able to coordinate their behavior with human behavior, which includes an aptitude to synchronize movements with their owners.[25] This behavioral synchronization, acknowledged to exist in diverse biological taxa,[26] has only recently garnered increased focus in studies concerning human–companion animal interactions. Social cognition, learning, and affiliation are believed to play roles in fostering behavioral synchronization between dogs and their owners.[27] Within a guide dog team, this shared attentiveness and alignment with each other's behaviors result in signals being received from the other organism as reflections of both the organism's condition and the surrounding environment concurrently.

Consequently, to make well-informed decisions regarding one's own future actions, it becomes imperative to consider the state of the other organism when interpreting the environmental cues they provide. Unlike electronic devices that often possess binary "on" and "off" states, the variability in the condition of the organism generating the signs can potentially undermine the reliability of the transmitted information. Paradoxically, however, the capability to accommodate this variability forms the foundation for emotional connections and trust between the handler and the dog. This in turn facilitates the cultivation of effective cooperation and the synchronization of movements.

There's also the potential for information from another subject to contradict information obtained directly from the environment. In such instances, one of the subjects can take the lead to benefit the team. For instance, a couple of my informants have shared stories where their dogs prevented them from falling or stumbling into obstacles by disregarding the "Forward" command and positioning themselves in the handler's path. However, there have also been instances where handlers didn't trust their dogs' signals, leading them into trouble. In this regard, the dog communicates information that's already contextualized, although the possibility remains that this contextualization is primarily from the canine perspective. Given that two organisms are moving together in a team, guide dogs must operate with a different understanding of the "body plan" and movement speed compared to what's programmed into the software of electronic devices, which typically consider only human morphology and behavior.

While both electronic devices and guide dog–assisted movement involve feedback loops between the device and user or the dog and handler, the latter is notably less linear and more dynamically unfolding. In a study investigating the cooperation of guide dogs and their handlers, Naderi and colleagues reported a to-and-fro initiation of activities between the two guide dog team members.[28] This implies that the handler and the dog alternately assume roles as information providers and receivers, and which subject takes on which role at a given moment is contingent on the specific circumstances. Similarly a comparable switching of functions between receiver and transmitter could be noted in the usage of electronic mobility aids. These aids begin emitting specific signals only after certain actions have been undertaken by the recipient (for example, when the person approaches a detected object). However, the sequence of reception and action is predetermined by the device's software, rendering the device's functioning foreseeable and reliable. Conversely, the role-switching between guide dog team members enables dynamic responses to the environment, taking into consideration the positioning of the organisms relative to their surroundings.

Both electronic devices and guide dogs convey environmental information through perceptible signals for the handler. However, while in the case of electronic devices the device itself ideally fades into the background of the user's attention (i.e., it is not an object of conscious thought), the efficacy of a guide dog relies on the continuous focus on the aid itself, concomitant with the dog's role as an extension of the human's perception. Gregory Bateson has employed the analogy of a white cane and a blind person to illustrate how specific perceptual tools become conduits of communication.[29] Although the guide dog and blind person together form a cohesive unit that orchestrates the flow of perceptual information through mutual coordination, this perception is unattainable without the communication that guides and molds the perceptual information. Through communication with the dog, the handler directs the dog's attention and perception, while through communication with the handler, the dog imparts environmental information that would otherwise be beyond the handler's reach. Consequently, even though the functioning of electronic travel aids could potentially be explained using models of extended cognition and perception, these models fall short in elucidating the cocreation of meaning between different organisms.

THE SEMIOTIC CHALLENGES OF MOBILITY AIDS

Reviews of various mobility aids consistently highlight that significant challenges associated with these devices are linked to the necessity of acquiring new sets of meanings and signaling patterns that these devices are designed to function with.[30] From a semiotic perspective, these issues can be categorized into concerns related to the signified, the sign vehicle, and the sign itself. The signified is the content or reference of the sign; the sign vehicle refers to the medium through which the meaning of the object is conveyed to the recipient; and the sign encompasses the signified, signifier, and the impact on the organism.

Regarding the signified aspect, divergent viewpoints emerge regarding the extent of information that devices should convey. Some advocate for devices offering intricate environmental details (such as distance, direction, and object surface characteristics), while others opt for simpler displays (such as only presenting the distance of obstacles).[31] Users' preferences may vary based on factors like the availability of other signals and the use of supplementary devices. Regarding sign vehicles, all electronic devices employ either auditory or tactile outputs. However, the challenge lies in aligning visually processed information with auditory outputs. Despite efforts to translate complex 3-D visual images into corresponding auditory representations, disparities in how these two types of coding are interpreted have hindered substantial success thus far. As for the signs themselves, the introduction of new meaning frameworks initially entails the need to acquire fresh symbolic relationships between objects and the carriers of these signs. Therefore if a user has never utilized sound pitch to gauge the distance of an object, they must learn the symbolic sign in place of the iconic or indexical[32] ones commonly understood by individuals with normal vision. Devices that leverage preexisting meaning frameworks include those employing human body topology, mapping it onto the environment. For instance, the Guleph project developed tactile gloves wherein the direction of obstacles was indicated by vibrations affecting the left or right fingers respectively. Similarly, devices with auditory outputs like the Sonic Pathfinder signal directions by emitting sounds either in the left or right ear.

The necessity of learning new signification schemes is also relevant for guide dog users, although this process is characterized by ongoing adaptations and readjustments. Nevertheless, guide dogs act as filters for superfluous information, making information overload or complexity typically less of a concern. Instead, the focus might shift to how to uphold the dog's role as an information provider. Guide dogs, being living organisms, partially rely on the bond formed through cohabitation with the handler. However, this interdependence entails handlers using motivational cues to elicit the dog's willingness to provide environmental information. Guiding constitutes the dog's role in specific situations, but this doesn't imply that all meanings pertinent beyond this context are suppressed. Although training and collaboration with the handler intend to bring guiding-relevant objects to the forefront and relegate those that hinder guiding to the background, the dog remains attuned to signals and objects specific to their species, breed, or individual character. For example, Labrador retrievers, the predominant breed for guide dogs these days, were historically bred as bird hunting dogs. Consequently, they often retain a heightened interest in birds even while fulfilling their guiding duties. As an Estonian guide dog user put it: "I can tell about my current dog, how she just like stretches out, that she is looking at that duck so much. And then I say 'Forward' and then she is still watching it, her neck turns twice, still watching, but 'Ok, I go forward.'" In such scenarios, simply repeating referential commands might not suffice to divert the animal's focus from the object. Instead, employing alternative signals such as an encouraging tone of voice, addressing the dog by name, or exhibiting assured behavior can prove more effective in redirecting the dog's attention back to guiding.

Guide dogs exhibit greater flexibility than electronic devices in forming and altering established sensorimotor connections. This adaptability arises from instantaneous learning, such as when the handler stumbles, indicating a misstep in guiding, as well as from recollections of prior situations tied to categorization and habituation. As machine learning advances, electronic devices may also enhance their flexibility in these aspects. However, the question of whether these devices will ultimately garner the same level of trust as dogs, which are esteemed for their capacity for independent yet interdependent decision-making, remains uncertain.

ENVIRONMENTAL AFFORDANCES

Different mobility aids utilized by individuals with visual impairments introduce varying requirements for environmental cues and offer distinct modes of access to the same surroundings. How can these dissimilarities be taken into account in environmental design? James Gibson's theory of affordances,[33] which centers on the alignment between an organism and the attributes of its environment, appears to offer valuable insights for achieving a better fit between individuals employing specific vision aids and the environments they navigate. However, despite the significance of Gibson's theories in architecture and design over the past decades,[34] several issues arising from the use of different mobility aids remain unaddressed in previous discussions.

For electronic obstacle detectors, the default programming entails considering all detected objects as hindrances to movement. Occasionally information about the size and shape of these objects is also transmitted, enabling the ascription of additional functions to them. However, conveying multiple functions of the same object concurrently encounters significant complications for two primary reasons. First, despite developers of electronic travel aids banking on the human brain's capacity to process intricate sound patterns, the need for extensive training to establish an equivalence between acoustic signals and intricate spatial information remains a prominent drawback of these devices.[35] In acoustic transmission the object is deconstructed into attributes, such as size, shape, and texture, each of which is systematically correlated with specific variables of the acoustic signal, such as pitch or tone. Disseminating information about the object via its individual attributes renders the interpretation of signals cognitively demanding, challenging habitual avenues of perception and understanding. As James Gibson succinctly put it: "The meaning is observed before the substance and surface, the color and form, are seen as such."[36] Electronic travel aids presuppose a perceptual process that operates inversely—individual attributes are initially presented to the senses and the meaning is subsequently sought. Second, even with the mastery of transmission codes, deducing an object's function from signals related to its form remains a complex task. Frequently, it's the context surrounding the object rather than the object itself that guides individuals in determining its function. How can an individual determine whether a stone functions as an impassable obstacle or one that can be easily stepped over or climbed? Beyond the object's inherent properties, its function hinges on the immediate physical environment (whether it's part of a stone fence or a pathway) and social affordances—what kind of actions are permissible given the presence of other people and the social context.

Physical and social affordances intricately define one another. For instance, furniture arrangement can prompt specific interactions, while implicit social agreements can render the presence of certain individuals acceptable in particular locations but not in others. In such cases, the guide dog's guiding role might be overshadowed by its species identity. The diverse interpretations attributed to the dog across different settings introduce uncertainty for guide dog users. They often find themselves unsure of the places where they are welcome with their dogs and where they are not. This dilemma is articulated in the words of a Swedish guide dog user: "An experience I have is that people think it's very nice and beautiful to see a guide dog walking together with me when I'm walking in the street, when I'm outside . . . But inside or if I'm in a situation, where people do not expected to find a dog—that could be like where I worked, in the hospital for example. Then I have always found that people are surprised, 'Oh, a dog' and they are a bit like, 'Mhmm, Ok now, let's see how we can solve this.'" As discussed earlier, the accessibility of contextual information appears to be just as vital for deducing the function of an object as the inherent attributes of the object itself. Concurrently, the requirement for contextual information presents a challenge to the theory of affordances, which posits that the shape of an object directly imparts its function to the user. Implementations of the

FIGURE 11.1A & B. The elements that serve as affordances for a person with normal vision but not for a blind person with a guide dog: (a) white line separating the pedestrian and bike lanes; (b) a curb stone that is even with the pavement. Photographs by Riin Magnus.

affordance principle in design have prompted the proposal that affordances should be perceptually self-explanatory for the user, effectively "naming" themselves without necessitating additional signals to convey the object's purpose.[37] However, underpinning such a proposition is an assumption about the specific capacities related to the specified affordances. In essence, affordances "name" themselves only to those who possess the capability to "hear" this "naming" (refer also to figs. 11.1a and b). Don Norman has introduced a distinction between affordances, which establish potential actions, and signifiers, which communicate where these actions should be executed.[38] While Norman's concept of "signifiers" pertains to specific design elements, the broader context of the object should also be examined in these terms: How could it be involved in the designation and specification of the particular affordance(s)? It would be valuable to not solely focus on the function represented by an object's shape, but rather to contemplate how to integrate it into an environmental context where its significance could be deduced, even if the intended affordance is not directly perceptible to the user. To facilitate the ease of attributing meaning, design should address functional scenarios by conveying complementary or even redundant information about objects within the environment. This approach could ensure that the designated affordance can be inferred with ease, enhancing the user's understanding of the object's purpose.

While both individuals using electronic travel aids and those relying on guide dogs encounter challenges in attributing functions to environmental objects, the nature of these challenges differs somewhat for each group. In the case of guide dog teams, the assignment of functional meanings is not solely the responsibility of the handler. Instead, it involves the active participation of another organism, the guide dog, prior to or in tandem with the user's interpretation. Due to differences in the Umwelten of canines and humans, the same objects may not necessarily serve identical purposes for these distinct organisms.[39] This alignment between capacities and environmental characteristics is a central tenet of the theory of affordances.[40] Nonetheless, human–guide dog teams represent a hybrid body configuration that doesn't perfectly mirror the capacities of each individual organism. Moreover, the capabilities of the interconnected organisms within a guide dog team are interdependent, and their consideration of each other's individual perceptual worlds benefits both entities. Given their role as guides, dogs are required to lead the handler to objects suitable for human use, not necessarily those that accommodate the dogs themselves. For instance, they should stop at sidewalk transitions that are relevant to humans, not dogs. This interplay exemplifies the dynamic relationship between the guide dog and the handler, where both organisms' Umwelten are harmonized for mutual advantage. Similarly, the individual should refrain from issuing commands to the dog that the dog is unable to carry out. Therefore the pivotal inquiry is not solely about how the human body aligns with environmental affordances, but rather whether the two organisms can collectively offer affordances to one another to effectively respond to the environmental affordances. This concept aligns with Gibson's notion of the "behavior affords behavior" principle.[41] In essence this principle underscores that the actions of one organism should guide the other organism toward appropriate modes of action within a given environment.

Although the dogs filter and interpret the environmental information before they transmit it to the user, their means of doing this are mainly limited to the tactile cues of the harness. Guide dog users must also acknowledge that the dog's interpretation of the environment might be rooted in a canine perspective rather than a human one. Consequently a similar contextuality of information is indispensable for both guide dog users and individuals employing electronic sensors to derive the accurate meaning of objects.

ACCESS BEYOND ASSISTANCE

As mentioned, accessibility encompasses more than just physical spaces; it is also intricately tied to societal norms regarding the presence of specific types of bodies within a given environment. Criticism has arisen against the notion that assistive technologies should solely aim to align with these societal norms, as such an approach can inadvertently perpetuate ableist values. Consequently it is essential to delve into the roles played by technical devices and guide dogs in shaping identities. This exploration should encompass not only how users establish connections with their mobility aids, but also how they interplay with the social and physical contexts they are situated within. In the words of Vasilis Galis, the emphasis should not solely be on assistive technologies aiding individuals in overcoming environmental obstacles. Instead, these technologies should be viewed as artifacts that encapsulate and manifest personal identities and capabilities.[42]

Accessibility is intricately linked to an individual's position within the complex web of social interactions. When considering technical devices, the main concerns revolve around their efficacy—that is, whether these devices are capable and efficient in accomplishing specific tasks. On the other hand, for guide dogs the questions at hand revolve around the societal norms that either permit or restrict the presence of nonhuman entities in particular environments. While the utilization of tools like a white cane or a GPS-equipped smartphone doesn't generally result in denial of entry to public transportation or restaurants, the same cannot be said for those reliant on guide dogs for assistance.[43] Such refusals based on social norms introduce a distinctive form of backlash unlike the challenges stemming from technical issues. In this scenario the rejection is fundamentally tied to the bond shared between two entities: one human and the other nonhuman. The direct rejection is aimed at the nonhuman companion, effectively reducing them to an "undesirable object." Indirectly this rejection also extends to the human counterpart, who to some extent relies on the capabilities of their nonhuman companion and for whom the "becoming with" another living being, as articulated by Donna Haraway,[44] holds consitutive importance in this context. Simultaneously these instances of rejection underscore an unwillingness to recognize that independence is often achieved through interdependence.

Rejection is a feeling that can arise not only in the realm of social interactions but also when confronting the implicit norms and values subtly woven into environmental design.[45] This design often serves to normalize specific body types. Sunaura Taylor encapsulates this idea: "We have become so accustomed to structures like steps and staircases that they appear almost inherent. Yet, curbs are no more innate than curb cuts, and flashing lights are no more inherent than beeping sounds."[46]

Technologies, when integrated into physical forms, establish expectations for a particular bodily presence, consequently driving the development of additional technologies aimed at aligning these bodies with these expectations. The assistive technologies and subjects are hence not mere devices for making Umwelten conform to the social and material affordances. They also function as means of interrogating the underlying realities wherein a specific entity must assume the role of providing assistance.

The persistence of animals alongside technical devices as mobility aids proves that the same goal of enhanced mobility can be reached through alternative means. When combined, the guide dog and the technical aid result in a hybrid form of aid contributing to the shared function, yet fundamental differences between these aids require a heightened attention on the part of the user so that one type of aid is not mistaken for the other. The dog is able to shift between independent and dependent decision-making, depending on whether the handler's instructions correspond to the environmental situation or not. Yet the handler has to seek for the same kind of correspondence between the dog's behavior and environmental cues. Electronic devices provide confidence that once a certain cue is there, its presence will be transmitted to the user, but the specification of its meaning is reached in combination with further environmental information. The seamless movement does not depend on the properties of the aid only, but how the aid can be attached to the habitual sensory pathways and the environmental affordances. Moreover, movement in the physical environment is embedded in the social environment, which further determines whether one is able to access the sought for social networks and to build one's identity on one's own terms. Individual human Umwelten that are grounded on the perceptual coding and interpretation of the environment are equally molded by the societal norms that significantly impact how one positions oneself within the social fabric. Interdependence is of constitutive importance for the self, but it becomes a problem if one is deprived of the possibilities to choose the interdependencies one wants to be a part of. In the words of Ingunn Moser: "Independence is not simply about disconnection, but also about the shifting out and replacement of some attachments (or dependencies) by others."[47]

Contextual design could help to specify the meanings of the environment in both cases by providing additional signifiers. Ideally context would help to ground the object in the field of meaning, both in the meanings of the physical object, and in the way the object is socially embedded, derivable from the attitudes and behavior of other people. The latter is often hard to comprehend even for people for whom the behavior of others is fully perceptible. As those cues cannot be inserted into the environment, special attention from other people is needed to make the social codes also accessible for those with different sensory abilities. In addition, the development of the animal and technological aids could consider each other's transformations (i.e., in which direction the electronic devices are developed and what needs to be taught to the dog). The field of multispecies studies stresses the need to consider the organisms in their co-becoming,[48] yet the same kind of co-formation could also be considered by the parallel transformations of technical devices and organisms. Considering the two as counterparts would perhaps allow the establishment of better symbiotic relations between them instead of developing them in frames of competition.

CHAPTER 12

THE HUMAN-CAMERA-ANIMAL TECHNOLOGICAL LOVE TRIANGLE

CONCEPCIÓN CORTÉS ZULUETA

Over time I have become aware that, in my animal studies academic practice, I usually write about trying to approach or become closer to other animals and their points of view, even about attempting to get inside their perceptions, perspectives, and worlds, to immerse in them. That is, I often think about other animals and about the ways to relate to them in terms of distance, even in terms of reducing or canceling that distance.[1] We tend to articulate human-animal relationships in terms of distance, in terms of modifying the distance between humans and other animals. Of course, other kinds of relationships and interactions also have spatial implications. However, distance seems to be playing a very relevant and significant role in how humans conceptualize their relationships with other animals, how we deal with them from our human side.

Then, if we add technology into this human-animal distance mix, we get a triangle instead of just a line. Analyzing this triangle becomes particularly interesting when the devices involved are cameras, when we deal with a human-animal-camera triad. Nowadays there appears to be a visual and technological love triangle between humans, cameras and other animals, widely broadcast through the internet and its social networks. It is remarkably fruitful to consider the articulation of these three elements—humans, animals, and cameras—regarding their positions and corresponding distances. Depending on the occasion, the distances and relationships among the elements of this triangle change, as the three of them try either to approach, chase or flee from one another—since, as tends to happen with any *love* triangle, they are not fueled solely by love. This chapter intends to address this human-camera-animal triangle and the different kinds of distances, and closeness, it entails, regarding its confounded physical, emotional, metaphorical, and optical dimensions and their implications, as well as the changes and variations they have experienced.

The prevalent initial attitude on the human side today seems to be to get as close as possible to certain other animals, including the wild ones that in the past were kept at a safe distance. At times this closeness is articulated in a physical sense, at other times in a conceptual or metaphorical one. Through their small size, innovations, and availability, even mobility and autonomy, cameras act as novel proxies between humans and other animals. Contemporary cameras can exercise multiple kinds of technological agency and can be mounted on drones and fly or swim toward animals, be attached to them and follow their movements as critter-cams, be disguised or act as traps to offer intimate perspectives that disregard animal privacy,[2] and be part of smartphones that are carried everywhere. On the whole, humans seem to be turning to cameras as an orthopedic[3] and technological cushion to soften the edges of human-animal interactions. As if it was becoming more and more difficult to conceive of or imagine animals without the assistance of these devices, in what implies an ocular-centric, vision-dominated mode of conceiving our interactions with other animals, as well as the animals themselves. However, the multiple ways in which cameras affect human-animal relationships don't always imply a smooth mediation.

Nonhuman animals in turn exhibit a wide range of agencies, of behaviors toward these devices and toward the humans behind them—behaviors that once in a while expose some of the absurd assumptions humans tend to hold about animals, or about cameras—for instance, when instead of ignoring it, a seagull steals a camera. Or when a wild or unfamiliar animal reacts and then proceeds to chase scared-to-death humans who apparently didn't see it coming.

This human-camera-animal triangle can be used as a sort of template to think about how the physical and metaphorical as well as emotional dimensions of its sides, of its distances, overlap and influence each other.[4] First of all, distance is a relevant and obvious topic to consider with regard to how encounters between humans and other animals occur and develop. Physical distance—or the lack of it—makes possible or precludes contact, touch, and it also determines if humans and animals are safe from each other. When a camera is present, we also can register the variations and evolutions of the physical distance between the three elements of the triad, and we can interpret their meanings and implications. This includes taking emotions into account and how they impact the distance between humans, cameras, and animals, as well as their variations. If the distance is reduced, it may be due to curiosity, trust, or affection, or maybe even predation; it may remain the same because of indifference or increase owing to fear. Consequently, beside considering the physical and metaphorical, in this chapter I will develop how the presence of cameras and the emotions we feel when we wield these devices toward other animals have molded the ways in which we manage and negotiate human-camera-animal distances.

On the other hand, the presence of the camera also turns distance into a technical, optical matter, with certain issues to consider regarding representation. In order to film or to photograph animals, it is of paramount relevance in which position and at what distance the camera is placed, which lenses and strategies are used, if the device is hidden or disguised, or if animals are somehow baited or lured, among other considerations. Likewise it is significant how the factual and visually recorded distance between humans, animals, and cameras is presented, and represented, both in the resulting images and in how those images are announced, broadcast, narrated, or advertised later on. All of this modulates our perception of physical distance introducing metaphorical and emotional aspects. On many occasions, in both the footage of wildlife films or online videos and in the ways in which that footage is presented and represented, there is an insistence on reducing the distance between us and other animals. A reduction that is framed as exciting and attractive, and that combines physical, emotional, metaphorical, and optical dimensions. As well, the prevalence of the optical, and of the interactions and relationships with animals mediated and dominated either by cameras or by their images, has led to particular forms of thinking about nonhuman animals and about their worlds, of shaping our understandings in this regard, and it has perhaps also persuaded us that we can even immerse ourselves into those worlds, into those animals.

FROM "FREEZE, FLIGHT, OR FIGHT" TO "PICTURE, PET, AND/OR PLAY"

As most people know, the whale is not a savage animal, and any small craft is reasonably safe among a school of these leviathans, provided they are not molested. Occasionally, however, a whale goes on a rampage, and then even fair-sized vessels have to get out of the way. The coasting schooner Cecilia had a stirring encounter with an immense hump-back whale off the coast of Nova Scotia recently, says the New York Ledger. When the monster, which the captain declares was seventy feet long, first made its appearance, the crew paid no attention to it, but when it swam alongside the schooner and gave it an occasional bump, the frightened sailors held a consultation. The only weapon on board was a 32-caliber revolver, and the captain fired one shot from this at the whale, aiming at what he thought to be a vulnerable part. But the only effect of the bullet was to further irritate the whale, and it hit the schooner several slaps with its tail that made the masts shake. Then it began to dive under the schooner from side to side, keeping the sailors in perpetual terror for fear he would rise directly under the vessel and turn her keel uppermost. There was no use trying to escape by flight, as the whale tacked every time the schooner did, and was twice as swift and they did not dare to use the revolver again. So the schooner sailed on for two days and nights with the unwelcome visitor frisking around her and not a man dared close his eyes. But the great animal was either merciful or ignorant of its strength. At any rate, it finally dropped astern after giving the Cecilia two terrible whacks that nearly capsized her.[5]

The physiologist Walter Bradford Cannon introduced the expression "Fight or Flight" around the 1920s to refer to the response behaviors that emerge to a perceived threat.[6] More recently, a third F is often added to these terms, that of "Freeze," whose relevance has lately been underscored against that of the other two, which now and then do exchange their positions.[7] Thus we end with Freeze, Flight, or Fight (FFF), a catchy phrase that has transcended the psychological realm and garnered certain presence and influence in popular culture. I won't focus here on the psychological processes and mechanisms per se, such as hormones, parts of the brain involved, or how these responses combine learned and innate elements. I am more interested in using the Freeze, Flight, or Fight topic to illustrate and contrast how our ways of dealing with animals have evolved and changed over time, in contexts that are now presided over and recorded by cameras.

In fact, it is not uncommon that, when explaining the FFF responses, the threat is imagined as starring an aggressive, predatory, or unknown animal.[8] The allegedly real situation described in the fragment of the *San Francisco Call* article above fits into this frame: a gigantic animal monster—in this case, a humpback whale—threatens a vessel and her crew, terrorizing captain and sailors for several days. Throughout the article the three FFF responses are mixed, appear, and reappear. They are advised, described, deliberated, evaluated, and attempted; they fail and succeed, in different manners and degrees.

The piece opens with a sort of recommendation. Most of the time, if you leave whales alone and unmolested, if you Freeze, they will leave you alone too.[9] However, at times "a whale goes on a rampage," and hence the best option for the vessel is Flight. After the whale bumps the schooner, the frightened sailors consult with each other and decide instead to Fight, firing the .32 caliber revolver. But the captain's shot causes further slaps and shakes—maybe a bit of Fight on the side of the whale too, but not too much, or the *Cecilia* might have been sunk. So, the Fight option fails, and the Flight response is attempted, but it is also unsuccessful because the whale is too swift. The only choice left is to wait—that is, taking no actions beyond sailing along their route—until the whale departs.

Overall the closeness to the point of contact between the humpback whale and the *Cecilia* is dominated by the fear of the sailors, who would prefer a little more distance, to flee themselves or to force the whale to flee through Fight. From our twenty-first-century standpoint, the decision to shoot the whale with a revolver is striking and harsh, maybe even a bit comically—or tragically—absurd. We don't really know about the emotions of the humpback whale, but the cetacean's behavior, or perhaps the account of the sailors' interpretation of that behavior, seems odd.[10] Humpback whale attacks do not appear real, from our knowledge and perspectives.[11] Indeed, it is far more common to read about humpback whales protecting other animals—humans included—from attack by orcas or sharks.[12] We are more familiar with another kind of whale-human encounter, usually captured by a multiplicity of cameras.

We click "▶" and the Earth globe that is seen on the screen becomes bigger and bigger as we loom over the region pointed to by a yellow arrow, a coastal area of the North American peninsula of Baja California, Mexico.[13] The name of the place, Laguna San Ignacio, appears next to a geolocation marker and a tiny picture of a gray whale, situated over a long and narrow lagoon. The image opens a bit awkwardly, as if behind a theater curtain, and reveals the turquoise surface of the sea where a dark shadow, barely discernable, begins to take shape. A gray whale emerges, sprays a plume of water, and approaches the hull of a boat, among exclamations of "awww" and "ohhh."

Then a series of takes and cuts follows, showing the interactions between two gray whales—a baby and a mother—and the humans on the boat. In the first one, the rostrum of the baby whale comes straight onto the boat's gunwale as the baby raises his or her head, as if examining the passengers. Indeed, the baby gray whale comes so close that the humans are able to pet, even to kiss the textured skin, already covered with incrustations. All of it while it sounds like a whale-watching guide enthusiastically shouts: "Touch them, touch them! Kiss them, kiss them, kiss them! Kiss the baby whale!" Afterward other approximations and contacts ensue, combined with shots of the revolutions of baby and mother around the boat, apparently taken from the same spot in the boat.

The passengers heartily laugh when the baby whale sprays them. The whole video is punctuated by other joyous onomatopoeia and expressions of wonder and pleasure, along with a few instances where the chirps of the whales are also heard. The human hands are divided either between holding cameras—included the one that filmed the images we are watching— smartphones on selfie sticks, and other similar contraptions, or trying to reach out to touch the whales, to reduce and cancel the physical distance. A woman clicks her tongue toward the baby while she bends over the water and splashes its surface trying to get the cetacean's attention, calling "hier, komm!" (here, come), as she probably would to a companion animal. Then she grabs the end of the whale's head with both hands, plants several kisses on it cheered by the guide ("beso, beso, beso!" [kiss, kiss, kiss]) and continues to pet the animal.[14] Finally, as a sort of farewell, we see how the tail of the mother waves and disappears into the sea, while the footage dissolves in a fade to black. All indications are that it was shot and put together by one of the tourists in the group, and the title chosen for the video—"Watching, Touching and Kissing Whales in Laguna San Igna-

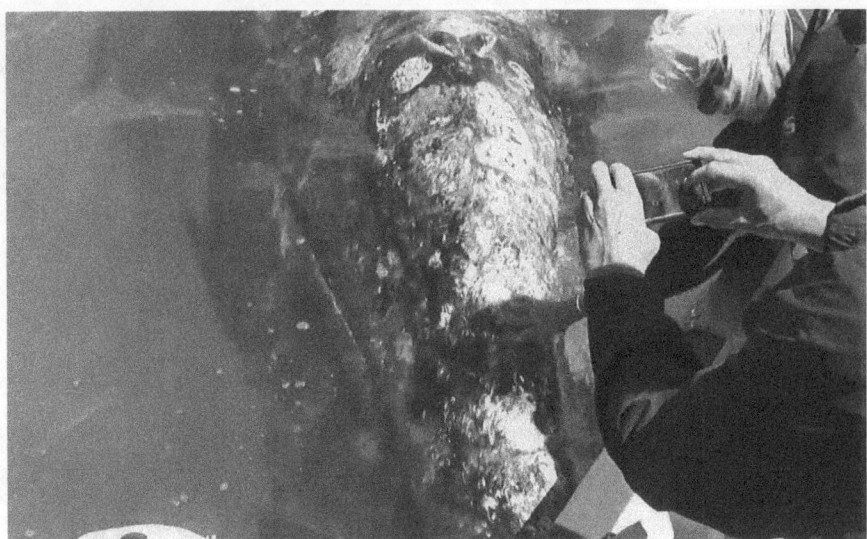

FIGURE 12.1. Screenshot from the YouTube video *Watching, Touching and Kissing Whales in Laguna San Ignacio / Mexico*, by Reinhard Döllner. *Source:* https://www.youtube.com/watch?v=I8ecwgaPMks, 2016. Courtesy of Reinhard Döllner.

cio / Mexico"—sums it all up, emphasizing both the visual component of the interaction and a cancellation of the physical distance tinged with affectionate emotions.

Therefore, there is a significant contrast between the 1891 *San Francisco Call* article describing an encounter between the sailors of the *Cecilia* and a humpback whale, and the 2016 video registering an encounter between the passengers of a whale-watching boat and two gray whales in Laguna San Ignacio. First, the main emotions underlying each situation are totally different.[15] The fear, even terror, experienced by the sailors is in direct opposition to the affection and the visible and audible pleasure emphatically expressed by the whale watchers. Both emotions impact the responses that emerged and developed during the encounters and how the physical distance between humans and whales was negotiated, increased or reduced. The sailors fluctuated between Fight and Flight options, and in the end they had to endure a tense wait; the whale watchers look forward to coming as close as possible to and touching the cetaceans in an amiable and playful atmosphere. The latter don't feel afraid when the whales dive around the boat; they are not thinking about how it can be capsized by the enormous animals, as did the *Cecilia*'s sailors, even though it seems probable that their boat is smaller than the schooner, and maybe smaller than the mother gray whale too.

In addition, in the whale watchers' case the contemporary multiplication and omnipresence of cameras, together with the prevalence of the visual, virtual,

FIGURE 12.2. Screenshot from the YouTube video *Watching, Touching and Kissing Whales in Laguna San Ignacio / Mexico,* by Reinhard Döllner. *Source:* https://www.youtube.com/watch?v=I8ecwgaPMks, 2016. Courtesy of Reinhard Döllner.

and digital context they foster, seems to enhance the already mentioned impulse to closeness, to reducing the distance between humans and certain encountered animals, like these wild gray whales. It is as if, even when we are confronting wild, potentially dangerous, or unknown animals, we may have replaced the FFF responses that could have been the reactions of the nineteenth-century sailors, with Picture, Pet, and/or Play (PPP) ones.

As thousands of photographs and films show, including the Laguna San Ignacio video, when a significant percentage of people encounter an animal these days, instead of freezing and being cautious, they try to digitally "freeze" that animal by taking a photograph or a video (Picture). In order to photograph or interact with the animal, instead of running away humans often try to come closer, and even to touch, the creature, assuming it has to be amiable to their advances (Pet)—as shown, for example, by the hands divided between holding cameras and reaching toward the gray whales. The schooner's captain wielded and shot the revolver, driven by fear, but here the firearm is substituted by this multiplicity of cameras that picture and are pictured, that film and are filmed. Susan Sontag wrote, "When we are afraid, we shoot. But when we are nostalgic, we take pictures."[16] The emotion present in the video wouldn't be nostalgia, but one akin to a pleasant euphoria, derived from that sought for and found closeness with the whales and associated with an accelerated context where it tends to be more relevant to digitally capture in order to immediately show and to communicate than to reflect on those pictures in the pages of a paper album, to contemplate them for years and years to come.[17] On the whole, it is as if those persons are taking for granted that the encounter and interaction has to be positive, pleasurable, and playful for all parties, far from a fight and more like play (Play). Thus instead of Freezing, Fleeing, or Fighting, if they engage in any physical exchange that implies contact, they presume it is Play and that they are going to be able to Picture and possibly Pet the animal. At least they tend to assume so at first, from their human side. This attitude is probably determined by how many animals tend to be presented and amiably framed, and cushioned, by visual culture, and how the impulse to reduce emotional and physical distance between us and them is encouraged, frequently through the adoption of metaphorical and optical frameworks.

In this preferentially visual and virtual context the different dimensions of distance between humans, animals and cameras (physical, emotional, metaphorical, optical), appear mixed and confused in a particular manner, which condition and influence each other. Affectionate emotions for other animals, encouraged by certain manners of representing and portraying them, make humans feel already close to those animals in an emotional way. This decreased emotional distance is then expressed in metaphorical terms, terms that either take for granted this human-animal closeness, this diminished distance, or look forward to reducing it even more. This reduction is also understood and carried out in a physical sense, not just in an emotional-metaphorical one. Therefore all these dimensions are conceptually confounded, and there is an impulse to translate the emotional and metaphorical closeness into actual physical circumstances—even to physical contact with those animals—to turn a decreased emotional and metaphorical distance into a reduced physical one.

Furthermore, in the human-animal-camera triangle there is a technical-optical dimension added, that makes people want to Picture how they attempt to Pet and Play. Driven by emotions and metaphors, more humans approach animals with cameras, and not just *through* cameras and their images, looking forward to canceling the physical and optical distance between them. Filming and taking pictures of animals, and then releasing and disseminating those images, used to be more of an operation for professional photographers. Now it is accessible to greater numbers of people that in a way can "go pro," as suggests one of the more popular camera brands, or that readily take out their smartphones and start shooting and sharing pictures or videos of those other animals they encounter. Close cameras register each episode that is then presented and broadcast in metaphorical terms of human-animal emotional communion and proximity. This is done in a manner that tends to favor specific optics, rhetoric, and aesthetics and that generally and mostly seems to reinforce the cycle above, in addition to imposing a predominantly visual mode to approaching other animals.

Some of the implications of this impulse to physical and optical closeness are that it can, and does, cause problems. First of all, more often than not this human attitude disregards how animals feel about the interaction, and imposes on them the involved humans' viewpoints and conceptions of the situation. A somewhat funny side effect of this concerns human expectations about animal behavior to cameras. Animals do exhibit a wide range of behaviors toward these devices, and some of them expose the absurd assumptions humans hold both about animals and cameras—like when a chimpanzee knocks down a drone with a stick, or a seagull steals a camera from the humans who were feeding this bird in front of it, in both cases, rupturing the assumption that a common human stance about cameras is also shared across the animal realm ("act 'natural,' as if the camera is not there!").

However, other consequences are much more damaging, either for the humans, the other animals, or both. Sadly, the more extreme cases involve situations such as baby dolphins dying while people take selfies with them, keeping them out of the water.[18] Several tourists have also died due to being too close, and too confident, when taking pictures around African elephants or hippos.[19] Similarly, there is a widely broadcast case, and a viral YouTube video, that involved a sea lion dragging a little girl under the water, probably by mistake, thinking that she or her dress was food. The video pictures the animal being fed once and then waiting for more while someone playfully mocks a feeding gesture toward the creature, inviting him[20] to come closer, as if trying to touch him. Next, we can hear a voice saying, "It's so cute." The sea lion's first approach to the little girl is met with laughs, nobody flees, and the sea lion is again invited to come closer. Suddenly closer becomes too close, fear appears, and everyone flees. increasing their distance after getting the little girl out of the water. The incident prompted Andrew Trites, director of University of British Columbia's Marine Mammal Research Unit, to lament the behavior of people not treating wildlife "with proper respect" and distance, as if disapproving of this impulse to Picture, Pet, and Play, to come physically closer: "You keep your distance. Watch the animals, but let wildlife be wildlife."[21]

But the outcomes do not have to be so drastic to be harmful. Animals approached or chased by humans with cameras can experience stress, and this is not difficult to imagine when we read studies that show how anole lizards react to the sound of a camera's shutter as if it was the call of a predator.[22] That is, the fear and the Freeze, Flight, or Fight responses quite often still define the reactions of the encountered animals. Moreover, sound—travels farther underwater and can be more annoying and disrupting. Also the presence of fast-moving boats poses a considerable threat to whales and other marine animals. Because of this, the International Whaling Commission has a set of "General Principles for Whalewatching," which advises, among other things, to "allow the cetaceans to control the nature and duration of 'interactions,'" to "avoid sudden changes in speed, direction or noise" and to keep "appropriate angles and distances of approach."[23] Following these and other expert recommendations, several countries have introduced or tightened regulations concerning the minimum distances that have to be observed when watching cetaceans, in general around 100 meters for large animals.[24] This sensible measure, at times and in certain countries, can cause human quandaries when the ones coming physically closer, to stationary boats, are the whales; in a way, the Canadian government request to "move away and keep your distance" when "a whale approaches you in the water" is a renewed sort of human imposition on the animals, theoretically for what is believed to be their own good.[25]

AN ORTHOPEDIC AND TECHNOLOGICAL CUSHION TO SOFTEN THE EDGES OF HUMAN-ANIMAL INTERACTIONS

Historically, the changing human meanings, shaped by the confluence of our cultures and biologies, that we project onto nonhuman animals have impacted how we conceive of and perceive them, as animal studies is committed to show. And in recent years it seems that digital technologies, and especially cameras—together with the ways in which animal footage is presented and represented and frequently distributed and amplified via social networks—have been conforming to a renewed kind of filter, one interposed in our relationships with many other animals, generating a barrier and modifying perception.

Metaphorically, this filter acts like a "cushion," even an "orthopedic cushion." On the one hand, one of the main effects of this technological filter can be to soften the edges of our encounters and interactions with other animals, to make them look more amiable, easy-going, and warm—like some of the plugins or social network filters with which humans transform and manipulate the pictures and registers of their daily experiences. This softening is mainly focused on the appearance of those interactions from the human side, on how we perceive and frame them, and not necessarily on the actual interactions themselves. On the other hand, the expression "orthopedic cushion" feels appropriate because it is as if more and more humans need to turn to this cushion, filter, or mediation in order to relate with other animals, as if they could not renounce to the use of this aid in order to relate with other animals—or acknowledge its side effects and implications.

At the same time this technological and optical cushion appears to be linked to how and when humans notice and acknowledge the presence of other animals and the manner in which they deal with them, how they handle and manage human-animal encounters and interactions. Even in polluted or animal-unfriendly cities, birds perceived as plain and everyday peck in the sidewalks, insects roam around the corners, rats thrive in the sewers. But humans tend to disregard or deny their presence; they don't get much attention, maybe because they usually fall outside the kind of animals softened through the aid of the orthopedic cushion. Cameras provide a script to guide and lead human-animal interactions, a frame and a cushion to reduce the friction, to smooth the uncertainty and uneasiness that those encounters might cause, and to turn them more into Picture, Pet, and/or Play-like experiences. You wield the camera, and it is not necessary to do anything else aside from picturing or filming how you try to get physically closer, how you attempt to pet and play. There is no further need to understand or to communicate with those beings. If Susan Sontag portrayed photography as a manner to deal with and possess past times, and to experience a sense of ownership of spaces in which people felt insecure—like when they travel to unknown places[26]—the orthopedic cushion at work in the human-camera-animal triangle allows humans to feel they control and occupy a present together with certain animals. Animals are chosen, softened, and shaped according to certain parameters that, in theory at least, are then easily incorporated into our virtual communities, like close digital members of the family.[27]

The problem is that the mentioned script encompasses two different sides, roles, and parts. Humans are following one of them, but the other one is for the encountered animals. And although the Picture, Pet, and/or Play script tends not to acknowledge it, nonhuman animals have their own interests and agencies and may not comply with the part that was written for them by human expectations. On the one hand, everything appears to go smoothly in the Laguna San Ignacio video: the gray whales reduce the distance on their own, the whale watchers are able to Picture, Pet, and Play with them. However, the sea lion interaction ends up being a different story. At first, all seems well: the animal looks friendly, gets closer, and is praised for his cuteness. But then, unexpectedly, he cracks the PPP script with a bold, un-cushioned action, leaving behind a group of disconcerted humans resorting to Flight responses they were not prepared for.[28]

This bewilderment is not so surprising if we bear in mind that the use of cameras and the viewing of their images bring with them that cushioned, softened framework: a visual, ocular-centric one with which we are familiar and feel certain, which advocates reducing any kind of distance and confounds its diverse dimensions (physical, optical, emotional, metaphorical). After all, nowadays many people, most of the time, with maybe most species, tend to visually learn how to identify animals (domestic, common, or exotic) and to find out more about them and their behaviors through photographs, films, documentaries, online videos—instead of directly watching and observing those animals in their living spaces while bodily sharing that same environment with them. This implies that the relevance of the visual aspect is underscored and emphasized, together with its optical context, script, and results. It may sound paradoxical, considering the impulse to becoming physically closer and petting the animals, but what appears to matter the most is not the tactile experience in itself. Instead, what counts are the images picturing that moment, that record and confirm the achievement, seen as a sort of culmination and climax of the Picture, Pet, and Play script, of the cancellation of any kind of distance between humans and other animals. Many of these images are, in fact, instantly and globally shared and consumed as a present than is more present that the present time itself, obscuring and leaving many other experiences and perceptions behind.[29] All of this reinforces the bias to pay less attention to whatever is ignored by that predominantly visual script, such as smells, certain sounds, or the nuances of the body language that may be exposing and anticipating its fracture and failure.[30]

Nevertheless, this doesn't mean that the current technological, optical proxy we are used to cannot be a valid or valuable way of acquiring knowledge about other animals.[31] For instance, despite the fact that I have never seen a shark swimming in open seas while diving close to him or her, when I was reading a short story by Anton Chekhov featuring an underwater shark who "sank down upon its back, then it turned belly upwards,"[32] I instantly felt that something was wrong, since I have never seen—in wildlife films and similar contents—such a fish swimming upside down, "belly up." However, one of the downsides of this manner of acquiring knowledge about other animals appears to be that too often we are overconfident and convinced that we know more about those animals than we really do, while simultaneously remaining unaware of the distortions, limitations, and discourses that accompany those images, and that we usually assimilate those distortions and meanings together with the images themselves. It's an ocular-centric circumstance that reinforces the loop and that makes it more difficult to depart from the current human dependency on certain technologies—like cameras, their use, and their images—when trying to interact and understand other animals. Likewise, all of this may be contributing to diminishing the abilities we need to relate with animals in other ways: nonverbal communication, embodiment, calm observation, leaving space for perplexity and awe instead of taking everything for granted, and even preventing our imagining that those alternative ways exist.

FROM LANDSCAPE TO PLANETARY BUBBLES

Recently there seems to be a development in how we conceive of and visualize animals, linked with the previously mentioned impulse to coming close and canceling distance, to an extreme kind of transparency whose roots were, nonetheless, already present in certain seminal and early wildlife films like *The Cuckoo's Secret* (1922) or *The Private Life of the Gannets* (1934), as manifested by their voyeuristic titles.[33] These, as well as many other wildlife films that followed them, promised to lead human eyes, our eyes, to where they couldn't otherwise go, to make visible and transparent the invisible, to make known everything there is to know about certain animals. Lately this tendency to an optical and conceptual transparency has been hyperbolized, as well as soaked with emotion, an intense desire for empathy and intimacy, which reinforces the impulse to reduce different kinds of distances unfolding between human and nonhuman animals, be they optical, physical, emotional, or metaphorical.[34]

In accordance, wildlife films of the twenty-first century are often announced and advertised, even proclaimed, as experiences that will take you closer to animals than ever before, or that will manage to take you inside one of them, to make you one of those animals. That is the case with National Geographic's *Crittercam* series (2003), criticized by Donna Haraway. A series that urges you to "sit back and imagine you are taking a ride on the back of the world's greatest mammal, or seeing life from the point of view of a penguin," while also promising that it will take you "as close as you can get to the animal world."[35] Haraway notes that these promises are not fulfilled, and we remain humans watching confusing footage filmed through (critter-)cameras that have been glued to unwilling animals.

Another instance of wildlife films proclaiming a renovated approximation toward animals would be the spy trend initiated by the director and producer John Downer, inaugurated in 2002 with *Lions—Spy in the Den*, where the spy introduced among a litter of cubs was a disguised and mobile "Bouldercam" designed "to get a cub's eye view of their action-packed lives."[36] In subsequent films, like in the BBC's five-episode series *Spy in the Wild* (2017), the robotic spies evolved into animatronic cameras, into human-made devices that looked and moved like animals—hence condensing in themselves the three conceptual elements of the human-animal-camera triangle, an achievement that is addressed as a milestone further reducing the distance between those three elements. Moreover, Downer's films focus on revealing the worlds of other animals, physically, perceptually or emotionally. This closeness becomes a constant leitmotiv, an aim emphasized by the introduction to the series, which also offers an intimate and emotional view of animal lives while wondering if they could "be more like us than we ever believed possible,"[37] underscoring a kind of nearness that can be read in behavioral or even evolutionary terms. Downer himself, talking about cameras and animals' worlds, states that he has "always been obsessed about how can we get closer."[38]

In *Spy in the Wild* human audiences are persuaded to engage in this quest to reduce physical, emotional, and metaphorical distances with animals in specific ways, for example, with footage subtly replicating or alluding to the plot of certain internet memes and viral videos found in social media and internet platforms—like seagull (or other animal) stole my camera!–type of stories, or animals-being-jerks compilations, among others.[39] Namely these are situations where humans, animals, and cameras are closely knitted together, optically, emotionally, and metaphorically. Sometimes this footage is uploaded to YouTube, and ends up working as a viral video in itself, closing the loop and extending a shared rhetoric of transparency, intimacy, and closeness between conventional wildlife films and many online animal videos. They invite the viewer to be close to those animals, to join them, even to be one with them, thereby contributing to the cushioned and softened perception of human-animal relationships that accompanies Picture, Pet, and/or Play–like responses during physical encounters.

Accordingly, both the images resulting from those encounters and the ones included in many wildlife films participate in the same trends associated with the impulse to reduce any kind of distance between humans and the rest of animals. Both might be responding to a change in how we conceptualize and visualize nonhuman animals' perspectives, worlds, and surroundings, or may be causing or reinforcing this shift through their optical approaches and aesthetics. In short, the said change appears to be replacing the notion and image of landscape, unitary and continuous, with a myriad of individual, subjective, and mainly visual planetary bubbles, which fragment a more uniform panorama. Here, "landscape" is understood as an elongated background that more or less adopts the shape of a canvas or a window, or one that we perceive as such, a background in which living beings drift and unfold.[40]

The transition from landscape to planetary bubbles may be tracked in the evolution of wildlife films. In a way it is as if we are partially going back to what was prevalent before, because the current planetary bubbles recall the "symphony of bubbles" metaphor conceived by Jakob von Uexküll to describe the interactions between different *Umwelten*, perceptual and subjective animal worlds.[41] As Gregg Mitman delineates in *Reel Nature*, between the 1930s and the 1940s wildlife films shifted from being intimate portrayals of individual species through the use of shots like close-ups—an optical aesthetics that could be seen as in line with Uexküll's ideas—to underscore "the importance of a wide-angle perspective on all of nature." Later wildlife films opted for a much more panoramic vision, going beyond the intimate and the individual species in order to encompass "an entire landscape," and to embrace "an ecological aesthetic, one that provided a glimpse into the complexity of the interrelations among organisms and their environments."[42] However, the current tendency—again, shared by certain online videos and some conventional wildlife films—appears to be reversing that trend, at least in part. Beside actually getting physically closer to the animals and engaging the viewers through emotional appeals or metaphorical acclamations of closeness, the resulting impulse also has an impact in visual, optical aesthetics.

In this sense National Geographic's *Crittercam* promises the point of view of the animals deploying images that were filmed by cameras glued to them. Downer not only describes himself as obsessed with getting closer to animals, but for him this also implies a particular choice concerning optics that entails using a "very little long lens, what it used to be the main style of natural history." Since "to get inside the animals' worlds" he tends to "shoot wide angle, my

FIG. 12.3. Screenshot and promotional image from the wildlife film series *Spy in the Wild*, 2017–2020. *Source*: Courtesy of John Downer Productions.

preference, getting the cameras up close."[43] In fact, this is another variation of the wide angle, so close to the animals that it is not uncommon that they fill the foreground and conceal the background; at times the disguised and anima-tronic cameras are even touched, tapped, licked, shaken, and carried around, by those animals.

According to Downer's words, and choices, in wildlife films like *Spy in the Wild* there is still a background, a landscape, but it is not as central as it was before, and it gets displaced and distorted (both optically and metaphorically) because of the importance granted to individual animals as characters and subjects. Also, to their presumed perspectives, they are featured in a sort of circular or spherical composition, which brings to mind the expression I have chosen to use, "planetary bubbles." Most of those perspectives would be simultaneously close-ups and wide-angled fish-eye views with a sort of bubble-like and slightly distorted aesthetics, the kind of footage or images often generated by spy-cams, crittercams, GoPros, mobile phones, et cetera, which are recorded and disseminated by both filmmakers and regular users.

On the other hand, the transition from landscape to little planets, from windows to bubbles, in our manner of visualizing or conceiving animals, and the worlds and environments they live in, can be articulated in other terms, more abstract and conceptual. The landscape model, heir to the Renaissance perspective and to the pictorial genre of landscape, suited a society and visual culture in which wildlife films were professionally made and frequently, though not always, brought remote locations and exotic animals either to the rectangular screen of public cinemas or to the private, smaller screen of the living room's

TV set. In a way both kinds of screens, housed in prismatic buildings, acted as windows to the world and to the animals, fulfilling a function that could be considered similar to that of the real windows opening into those same buildings. These real windows were ideally imagined or intended to show nature or bits of nature from the neighboring landscapes, inhabited by creatures that now and then popped up here and there.

But these days we are surrounded by a multiplication of other kinds of windows, screens, many of them portable ones that come in any size and that most of the time not only reproduce images but are also able to record them.[44] Therefore all these screens, which in turn lodge plenty of virtual windows inside themselves, don't fit in buildings or in a strict landscape notion, though because of their shape and characteristics, they could be considered as an extension of this landscape model. Still, they depart from it because they are no longer housed inside while showing something outside. We move around and carry them with us—sometimes, they are carried away or grabbed by other animals too—and hence they can be everywhere. Therefore they are a very peculiar type of screen, of window, and surely not a fixed one. A high percentage of them, like mobile phones, tablets, and laptops, are intimately and tightly—in a metaphorical and emotional sense—attached to a certain subject, to a certain individual, and through its constant use, its reception and emission of images, it is as if this individual was continually generating his or her own visual bubble, a little planet radiating from the focal point that this particular subject is occupying at each given instant.

So maybe it is not that surprising that we might be extending this same planetary bubble model to the manner in which we conceive other animals as subjects, or how we visually imagine and represent those animals, and that this can be associated with the ways in which we try to approach them, to become intimate and close to them, physically, metaphorically, emotionally, and perceptually, and above all, visually. Even to immerse ourselves into them as if they really were little subjective planets, sensory bubbles inside which we could submerge to become one with other animals and to experience the surrounding environment through them. Somehow, this immersion would be the ultimate ocular-centric climax of an approximation process and impulse that, encouraged by the cushions that soften human-animal encounters through cameras and their images, confounds the different dimensions of distance that I have addressed in this chapter: an impulse to cancel distance and become closer and closer to other animals that favors human Picture, Pet, and/or Play responses instead of Freeze, Flight, or Fight ones, often regardless of what the animals themselves think or feel.

EPILOGUE

ENTANGLEMENTS AND VISIBILITIES

JENNY LEIGH SMITH

The thirteen essays in this volume highlight how humans and other animals share space, either how they have done so in the past or how they currently negotiate the terms of shared spaces. It is striking how many of these interactions are exploitative ones in which humans use animals as tools or unequal partners. Nevertheless, human-animal relationships are shifting, complex, and multidimensional. In paying attention to the partnerships, dependencies, and other forms of interspecies symbiosis animals and humans create, the authors of these essays vividly describe how human-animal relationships continuously help make and remake the world in which we live.

Humans, cows, dogs, eagles, Southern elephant seals, right whales, elk, penguins, and European wolves are just some of the charismatic megafauna with which this book is concerned. In this epilogue I revisit two central organizing ideas the editors and authors have introduced as a way of summarizing and reconsidering the important themes of this book.

The first of these is the theme of entanglement. Finn Arne Jørgensen and Dolly Jørgensen use this word to frame their introduction, and although entanglement can refer to a generic form of interrelation, a more precise definition will also be helpful. A second major theme explores ways of seeing. How do technologies, specifically technologies of visibility mediate the human-animal relationships described? The book's contributors variously name this phenomenon "ocularization," "visualization," or more prosaically "photography," but they mean something similar: sharing spaces with animals in the modern world has meant rendering these animals more legible—specifically more visually present—than ever before. This calls to mind James Scott's now-classic writings on the sight-centric modernizing impulses of state power and also John Berger's essays about power and visualization that focus on processes of dominance and submission, commodification and revaluing art and aesthetics.[1] These complementary themes of entanglement and visibility will guide this summary.

Entanglement suggests interconnectedness, but also disarray, tension, and asymmetry. The stories in this book highlight intersections between humans and animals, and locate animals in previously human-centric narratives. These chapters evaluate the power dynamics between species and describe how various technologies, from radio collars to aquarium glass to calorimeters mediate these relationships. As the editors note in the introduction, the philosopher of science Donna Haraway uses the term "entanglement" to explore the depth, pattern, and meaning of human-animal interactions.[2] Entanglements are ties that connect, bind, constrain, and evolve.

Philosophers and historians are borrowers; the term "entanglement" is borrowed from the field of quantum physics. To quantum physicists, entanglement is the phenomenon of different categories of subatomic particles behaving dependently and predictably even when no physical or chemical link can be perceived between them. Einstein famously called quantum entanglement "spooky interaction at a distance," and although he found the theory intriguing, he doubted its existence. Proving Einstein wrong and demonstrating that this kind of invisible influence could and indeed did occur was a major achievement of twentieth century physics. Quantum entanglement means that if you know something about one system, you know something about another system without observing it directly. The systems are linked, their fates intertwined in ways people do not yet understand but can now predict. Spooky interaction indeed. For Haraway and other social scientists, this portmanteau serves as a touchstone in their work that considers human-nature and human-other relations; invisible yet linked, predictable yet mysterious.

Revisiting these chapters with this more precise definition of entanglement yields fruitful insights. Look at humans, and you will know something about their relationships with animals. Look at animals, and humans come into sharper focus. Some of these relationships are clearly visible, others are more inscrutable, yet still predictable. In many of them, the technologies that mediate human-animal relationships flavor the nature and meaning of the entanglements we observe.

For example, Nicole Welk-Joerger's steer is planted firmly in a capitalist industrial economy where metabolism, respiration, micturition, and other emissions are monitored for the purpose of controlling costs and maximizing profits. Surplus capital is dedicated to bringing more carefully curated cattle into existence. Future stock will also be nourished, monitored, and consumed by the appetites of twentieth century meat lovers. In this scenario, to know the steer is to recognize that he is—by sex, breed, size, age, and growth rate—an intermediate step in a twentieth-century American quest for the optimally efficient unit of edible beef. Welk-Joerger posits her 1901 steer in a story of food anxieties, but this is also a story about food hubris. These putative improvements to animal agriculture are aligned with an endless, frictionless, and impossible ideal of growth. Welk-Joerger's steer has a story to tell about the origins and ambitions of this ideal and the economic logic that allowed bizarre cultural practices, such as locking an animal in a steel box for several days, to be interpreted as a rational form of scientific inquiry.

Tatsuya Mitsuda is also concerned with describing the ways that beef cattle and humans became entangled by a modernizing state. First Germany and then Japan colonized the mainland Chinese port city of Tsingdao in the early twentieth century. As in Welk-Joerger's chapter, Mitsuda focuses on how cattle production expanded as the territory modernized. Rather than focus on herd management techniques or feed innovations, Mitsuda emphasizes the role of animal health and disease prevention. Both colonial states made extending life, improving survivorship, and ensuring healthy, sanitary food supplies as major goals. The increasing mobility of both animals and people stymied these goals. New technologies, such as tests for brucellosis and refrigerated train cars meant that creating good sanitation and health protocols became increasingly more complicated.

What can we learn about German and Japanese colonizers from Tsingdao's cattle? First, that both sets of colonizers thought of themselves as somatically vulnerable in ways that set them apart from the local population. Colonists perceived beef to be safer to eat than pork, and cattle dispatched in European-style slaughterhouses could be rendered into a more appealing and hygienic product. Rather than relying, as the local population did, on experienced butchers to deliver the freshest cuts, colonizers trusted trained veterinary professionals to monitor the lives and deaths of Tsingdao's cattle to ensure the beef supply was clean and disease free; freshness was secondary. Tsingdao's beef cattle lived comparatively cosmopolitan lives, and died rapid, efficient deaths. From a twenty-first-century vantage point, Tsingdao's slaughterhouses and beef exporting practices seem prescient, but in early twentieth-century Tsingdao the rise of professionally managed slaughterhouses was unique and difficult to understand. The logic of colonial capital and the biopolitical priorities of colonizing powers were profoundly out of sync with local norms. Mitsuda's case study of Tsingdao admirably shows this disconnect.

How different is the animal-human experience of the nomadic pastoralists, as they trail their sundry "five muzzles" through the steppe and taiga of Inner Mongolia than that of the midwestern steer Welk-Joerger traced? By the end of the twentieth century these experiences are not very different. Should this surprise us? Perhaps not. Aurore Dumont provides thick descriptions of daily life in herding communities before and after industrial logic arrived on the steppe.

In this chapter it is Chinese communists who are the standard bearers of modernization. State officials encouraged pastoralists to enlarge herds, disaggregate mixed flocks, and to retain only the most profitable species and breeds. Officials adopted pronatalist husbandry policies and provisioned animals by introducing mechanized mowing machines to increase the winter hay supply. Herders were encouraged to settle in permanent communities, while at the same time roads, motorcycles, trucks, and camping vehicles became pervasive. Populations that had previously paced their movements to match their hoofed chattel rapidly evolved their patterns of travel and property ownership to adapt to the new possibilities and limitations of mechanical horsepower.

In Inner Mongolia, herd populations and their human companions encountered dramatic, state-sponsored changes. Many of these were the kinds of routine changes modernizing states have performed globally: simplification, scaling up, and acceleration of production. A unique correlation Dumont makes is that the arrival of heavy, hard-to-transport housing and other bulky possessions motivated herders to comply with state projects of modernization. As herders settled into fixed communities, they literally lost sight of their herds, as single-species herds and flocks were easier to manage periodically or even remotely than the mixed flocks they had previously tended.

The next two chapters deal with hunting: Karin Dirke is concerned with the rise of game management principles and new standards of animal welfare in Swedish hunting in the late nineteenth century, and Finn Arne Jørgensen focuses on the adoption of GPS radio collars for hunting dogs by Swedish moose hunters in the contemporary period. In both stories new technologies force hunting practices to adapt and evolve, yet the fundamental rationale of modern hunting perseveres: hunting practices are recreational, social, responsible, and culturally traditional.

Dirke focuses on hunting narratives that appeared in a Swedish-language sportsmen's magazine in the late nineteenth century. These are the first generation of hunters to have access to more deadly and accurate guns, and they also have more leisure time in which to embrace hunting as a sport. As hunting expands as a hobby in this era, Swedish prey are more vulnerable and more targeted than ever before. At the same time a new concern with animal welfare and an increasingly sentimental and romantic attitude to wild places and wild animals create new social tensions. A growing number of people have begun to view killing for sport as cruel and actively campaign against hunting. Dirke's history describes an important moment in animal histories, one in which humans begin to act as advocates for animals and to publicly express empathy for their suffering. This emergent shift in attitudes is a side effect of technological modernization.

Finn Arne Jørgensen's description as a participant observer in a Swedish moose hunt in 2018 contemplates the roles of technology, tracking, and hunting dogs as assistants and intermediaries in moose hunting groups. The participation of dogs in these hunts is long-standing, and the breeds of dogs used in the hunt are, likewise, traditionally limited to a regional few. Yet Swedish hunters have readily adopted various locative technologies, first radio collars and more recently GPS monitors, in order to help them trace their dogs and streamline the hunt. Jørgensen's fieldwork reveals that hunters don't worry that new technologies will alter the fundamental relationship of trust and delegation that exists between hunters and their dogs. Dogs are still expected to check in on their humans, to sound when they sight prey, and to safely track a wounded animal through the woods. In this instance, GPS acts as a noninvasive assistive technology, augmenting rather than replacing hunting dogs' skills.

What can a dog wearing a GPS collar and reflective hunting vest tell us about her human collaborators? Most strikingly, that hunting is a form of serious play.[3] Moose hunting may be a social recreational activity, but it is also big business. Training and caring for dogs, purchasing firearms and ammunition, taking time away from homes and workplaces, all for an uncertain reward, are small luxuries that add up. Nevertheless, moose hunting remains a popular pastime among a significant subset of Swedes. The enduring activity of moose hunting in spite of its poor return on investment is a testament to its powerful role as a central cultural activity.

The next two essays deal with wildlife management: Heta Lähdesmäki's article explores the history of using radio collars to monitor the ranges of European wolves in Finland, and Tuomas Räsänen, looks at care protection for white-tailed eagles along the Baltic Coast and the salvage work humans have done in order to partially restore their populations. While both essays are focused on wildlife protection, they offer different lessons about how human intervention in wilderness settings can transform the lives of wild species.

In the case of Lähdesmäki's wolves, the creatures are involuntary research subjects in a study that ultimately harms the wild wolf population. By closely tracking the wolves in Finland, wildlife management officials gathered important information about wolf spatial distribution, but a significant number of these subject-participants were injured or killed as a result of wearing tracking collars. There is a major "entanglement" insight embedded in this history: the wolves were harmed because scientists failed to correctly analyze the risks of the study. In order to proceed ethically, researchers must account for these risks when designing future studies.

To learn about the experience of radio-collared wolves in Finland between 1970 and 2000 is to understand that human researchers had conflicting research priorities as they sought to create a database of wolf behaviors and movements based on radio-collar signals. Wildlife managers hoped to use the study's data as justification to kill excess or off-range wolves. Field biologists wanted to expand basic knowledge about wolf movement and diet. Finally, conservationists wanted to tell personalized stories about wolves that could help build public empathy and respect for a maligned species. These conflicting priorities indicate the varied levels of compassion these different groups had toward the wolves.

Räsänen is also interested in compassion for wildlife. In the case of the white-tailed eagle, a stunning reversal in human attitudes to the animal emerged rapidly. First, an all-out effort to eradicate the eagle in the early twentieth century enjoyed great success. Then, in the post–World War II era, human-made toxins like DDT nearly finished the job. Finally, new conservationist values created an agenda to protect and recover the eagle, and this project enjoyed modest success that has continued into the contemporary era. In this case study to observe the fate of the white-tailed eagle is to learn, again, that humans are fickle and unreliable stewards with infinitely malleable attitudes toward nature. For these eagles the future may hold some sort of redemption; thanks to vigorous PR efforts, human disgust has turned to admiration of the eagles in a single generation. Stable, sustainable species recovery will take much longer.

Several essays target the historic gaze between humans and animals and focus specifically on several different forms of visibility. One of the major observations is that this gaze is not only unequal, but that technologies amplify and distort this inequality. Sometimes these amplifications and distortions are welcome; other times they become undesired distractions.

Ellen Arnold's chapter is also concerned with how new technologies of transit and communication helped turn penguins into mediated, distorted commodities in interwar America. The animals in question are twenty "leftover" penguins originally imported to enliven a temporary exposition about Richard Byrd's first Antarctica expedition. The penguins are precious cargo, but it is too costly to return them to their native habitats; instead they are put up for sale. Proceeds are earmarked for future Byrd trips. Letters of inquiry about purchasing or adopting Byrd's birds reflect an enthusiastic but naive American attitude toward charismatic megafauna. Byrd's penguins become the playthings of Manhattan elites and the residents of parks, winter carnivals, and zoos across the country. One becomes the mascot of a midwestern university. In Arnold's case study technologies of enhanced and broader mobility fuel an enthusiasm for exotic animals that are associated with a quintessentially modern voyage of discovery.

In Dolly Jørgensen's chapter contemporary theme park aquariums are entertaining, educational environments. One of their tasks is to make previously invisible or murky underwater worlds visible and even visually stunning to land-bound humans. Strong, lightweight acrylic barriers have made much larger aquarium exhibition tanks possible, but acrylic's plasticity also distorts and limits aquatic exhibits in new ways. A ten-million-gallon tank may impress, but aquarium designers must engage visitors in other ways. A surprisingly successful trend has been a digital turn: high resolution LED screens now complement the live-action experience at many marine parks. Visitors watch live and prerecorded footage of exhibits. They are appealing because, unlike their unmediated counterparts, on-screen animals are lively, in focus, and doing interesting things. In these exhibits witnessing aquatic life firsthand is less important than an unobstructed view of animals behaving spectacularly; visibility wins over verisimilitude. In Jørgensen's account contemporary aquatic exhibits expose the differences in perception, gaze, and display between terrestrial and aquatic worlds. Screens anthropomorphize aquatic animals because they amplify underwater visual realities and make them more accessible to humans. Like acrylic tanks, LED screens distort, but they also complement and enhance the live-action performances that aquatic theme parks host.

Charity Edwards and Amelia Hine focus on the experiences of wild marine mammals, specifically southern elephant seals, serving as underwater eyes. Rather than valuing animals as food, southern elephant seals are valued as biological probes, able to carry cameras and other sounding devices hundreds of meters below sea level to help locate new sources of rare minerals and other elements. The seafloor, uninhabited by and physically inaccessible to humans, can be surveyed by proxy by these well-adapted marine animals. The major insight these animal subjects offer is to expose the physical fragility and economic cunning of human collaborators that have developed such ruthless yet creative forms of animal exploitation.

In Riin Magnus's chapter on guide dogs and assistive technologies for the blind, trained animals trump various forms of electronic surveillance designed to help blind people navigate their environments more seamlessly. While the mobility aids are intended to increase the amount of environmental information available to blind people, one user described the feedback from obstacle detectors as "a salad of noises." Users prefer working with guide dogs as a team. Mutual attentiveness and the dog's ability to switch between following orders and giving directions make it a more flexible form of mobility aid that beeping sensors could not match. In this case the entanglement between humans and animals reveals a reciprocal relationship; the dogs are capable of building an organic and intuitive rapport with their human interlocutors that contemporary visualization technologies cannot match.

Finally, Concepción Cortés Zulueta's chapter takes technology-as-mediator eriously and considers human-animal relationships by taking a closer look at a technology that often mediates their connection: the camera. In Cortés Zulueta's account cameras have fundamentally changed the relationship between people and wild animals. Humans imagine pets and playmates where they might once have seen physical threats. Cameras make wild animals more approachable and more relatable. This emerging sentimentality tempts people to embrace rather than eschew wild animals, with potentially negative consequences. Here, as with Dolly Jørgensen's chapter, there is a fundamental failure of imagination on the part of humans as they consider their responsibilities toward the animals with which they are entangled. Animals in this human-camera-animal triangle are misrepresented, their identities distorted.

VISIBILITY

The discussions of cameras, displays, and visual assistive technologies at the end of this book provide an opportunity to segue into the second theme that connects these thirteen essays. Beyond the entanglement of humans with animals, there is also a theme of visibility. Some chapters address this visibility explicitly, but this same motif pops up in many of the others as well. Metaphors of "seeing" and "making visible" are so closely tied to how we express our understanding and comprehension of the world around us that it is almost impossible to uncouple sight from knowledge. As John Berger wrote in 1972, the relationship between what we see and what we know is never settled, and the way we see things is affected by what we know.[4]

These robust links between seeing and knowledge have been amplified by modernization, especially forces of modernization that have expanded the catalog of what is seeable. Advances in optics and microscopy allow us to see increasingly smaller objects, and new technologies like radar and spectrophotomics have made previously invisible or unknown natural forces visible in meaningful ways. While not all of these advances in visualization technologies relate directly to the evolving relationships between humans and other animals, interspecies lives are so intermingled that many of them do.

There are three genres of visibility in *Sharing Spaces*. First is the visibility of tracking, such as that used on the Bear 71 featured in the introduction. Tracking Bear 71 allowed filmmakers and scientists to create a coherent and eminently watchable video biography of a Canadian grizzly bear. The practice of tracking—precisely knowing where an object is in space and time—derives from military and hunting technologies but has been embraced by a much wider audience today. A second form of visibility is that of accountability. Who or what can and should be measured? The excrement of cattle? A population of endangered eagles or "rogue" wolves? Modern states are often motivated to make visible the elements of society they want to control or to give the appearance of being able to control. Finally, there is the visibility of spectacle, perfectly encapsulated by the rising celebrity of salmon-catching bears in Katmai National Park. Animals on display are visible, judged, commodified, circumscribed, and valued in ways unique to modern times and modern entanglements. Consuming and judging these displays is work that humans do that reorders and revalues their worlds and their interspecies relationships in important ways that have changed significantly.

Tracking techniques and technologies make the movements of animals more visible. Some animals, like the hunting dogs in Finn Arne Jørgensen's chapter, are tracked by their owners to help them locate their quarry and to reassure the owners their dogs are safe and not roaming too freely. The southern elephant seals in Edwards and Hine's chapter are tracked for the purposes of triangulation. The cameras and other probes attached to the animals yield meaningful information only if prospectors can pinpoint the locations on the seabed where cameras and probes indicate that rare minerals are likely to occur. In this case visibility is tied to an emerging technique in resource extraction where previously invisible and inaccessible parts of the planet are targeted as new sites of production. Finally, tracking was one technology among several that were employed in the successful care protection approach Sweden and Finland adopted in the early 1970s to try to rebuild white-tailed eagle populations. Conservationists used airplanes to conduct aerial surveys to hunt for previously unidentified eagle nests, and they climbed trees to band nestlings. These labor-intensive tracking techniques helped conservationists monitor living birds and contributed to understanding eagle deaths. Eagles lived and nested in remote areas, so intensifying surveillance was expensive and complicated, but the success of the larger project seems to indicate this hard work paid off.

Next is the visibility of accountability, in which making information visible by recording it, charting it, and calculating its change over time and standard variation makes new forms of knowledge possible and changes the relationship between the observer and the observed in the process. Accounts, inventories, and surveys are never neutral technologies. For example, Matthew Edney has argued that by creating maps of colonial India, colonial administrators were able to use them to claim more authority and control over the region.[5] Observing and categorizing animals are activities that demonstrates the influence culture has on scientific practices, one that could incorporate both scientific and laymen's values and this "consistently inconsistent practice."[6]

Welk-Joerger's calorimeter is the stand-out example of this practice of accounting for animal lives in increasingly quantifiable and precise ways. Inputs and outputs of air, water, and an optimized and streamlined diet were all painstakingly calculated in an effort to extract ever more value from the American beef steer. In the 1970s white-tailed eagles were measured as a way to increase visibility and knowledge about how they had become endangered. Eagle populations were measured at both the macro and micro levels; an extensive census of birds and fledglings tallied the remaining birds, but scientists also measured the toxic burden of PCBs and pesticides like DDT in eagle carcasses and nonviable eggs. Emerging techniques in gas chromatography and mass spectrometry were refined and repurposed to allow toxicologists and ornithologists to better understand how industrial contamination threatened eagle populations. Finally, measurement as a form of visibility plays a key role in the "intra-action" of wolves and people that Lähdesmäki describes in chapter 6. Of the book's authors, Lähdesmäki is the most explicit in linking culturally informed methods of tracking and accounting with the kind of knowledge that can be produced.

The third and final form of visibility that forms a common theme in many of this volume's chapters in this volume is that of display. Animals as human spectacle have a long history, but they are also a topic of contemporary enduring interest. The fat bears of Katmai gain thousands of social media followers every year, and while the economic costs of opening and maintaining public aquariums have risen sharply in recent years, many still rake in profits. However, beyond these self-conscious and carefully curated forms of display, the two chapters on hunting also describe evolving norms of hunting practices that clamor for a more direct experience of the hunt, with new technologies—in one case satellite technologies, in the other affordable media publications directed at the armchair enthusiast—to turn hunting into a sport that can be witnessed even from a distance.

Ultimately, displays of animals, whether in the past or in contemporary times, are all worth considering in terms of the kinds of technologies that mediate the experiences between observer and observed. Technology allows us, in the words of this volume's contributor, Cortés Zulueta, to "get inside the animals' worlds," to get closer to animals. Cortés Zulueta also points out that this proximity, the very see-ability of animals consistently leads humans to mistake the social and power relations between people and animals. The spaces that humans and animals share today are more crowded, polluted, and fraught than at any previous point, yet there is still much to learn from taking a closer look at past and present human-animal interactions to try to locate the best possible way forward, learning from both the missteps and promising starts that have been made.

NOTES

INTRODUCTION

1. Jeremy Mendes and Leanne Allison, "Bear 71," *National Film Board of Canada*, 2012, http://bear71.nfb.ca/.

2. Tina Loo, "The Bow Valley and 'People' without a History," *NiCHE*, January 10, 2018, http://niche-canada.org/2018/01/10/the-bow-valley-and-people-without-a-history/.

3. Colleen Campbell and Tina Loo, "Making Tracks: A Grizzly and Entangled History," in *Traces of the Animal Past: Methodological Challenges in Animal History*, ed. Jennifer Bonnell and Sean Kheraj (Calgary, AB: University of Calgary Press, 2022); open access edition, accessed November 14, 2022, https://ucp.manifoldapp.org/read/traces-of-the-animal-past/section/11e6f06b-fd4f-49fe-9ad5-fd93c977006e.

4. As demonstrated in Aaron Koblin's visualization of air traffic over North America: "Flight Patterns," accessed November 14, 2022, http://www.aaronkoblin.com/project/flight-patterns/.

5. Finn Arne Jørgensen, "The Internet Is Obsessed with a Video Feed of Bears Eating Salmon," *The Atlantic*, August 22, 2016, https://www.theatlantic.com/technology/archive/2016/08/bear-cam/495638/.

6. Dolly Jørgensen, *Recovering Lost Species in the Modern Age: Histories of Longing and Belonging* (Cambridge, MA: MIT Press, 2019).

7. Roger Thompson, *No Word for Wilderness: Italy's Grizzlies and the Race to Save the Rarest Bears on Earth* (Ashland, OR: Ashland Creek Press, 2018).

8. Etienne Benson, *Wired Wilderness: Technologies of Tracking and the Making of Modern Wildlife* (Baltimore: Johns Hopkins University Press, 2010).

9. The workshop "Closing the Gap: How Technology Changes Spatial Relation-

ships between Humans and Animals" was funded by the Nordic Centre in Shanghai at Fudan University and organized by Finn Arne Jørgensen and Dolly Jørgensen.

10. Thom Van Dooren, Eben Kirksey, and Ursula Münster, "Multispecies Studies: Cultivating Arts of Attentiveness," *Environmental Humanities* 8, no. 1 (2016): 3.

11. Bruno Latour, *We Have Never Been Modern*, trans. Catherine Porter (Cambridge, MA: Harvard University Press, 1993).

12. Donna Haraway, *When Species Meet* (Minneapolis: University of Minnesota Press, 2008).

13. For examples of scholarship untangling the mess of relations with domesticated species, see the essays in Susan Schrepfer and Philip Scranton, eds., *Industrializing Organisms: Introducing Evolutionary History* (New York: Routledge, 2004); Heather Anne Swanson, Marianne Elisabeth Lien, and Gro B. Ween, eds., *Domestication Gone Wild: Politics and Practices of Multispecies Relations* (Durham, NC: Duke University Press, 2018).

14. See Van Dooren, Kirksey, and Münster, "Multispecies Studies."

15. Susan Nance, "Introduction," in *The Historical Animal*, ed. Nance (Syracuse, NY: Syracuse University Press, 2015), 5.

16. Donna Haraway, *Staying with the Trouble: Making Kin in the Cthulhucene* (Durham, NC: Duke University Press, 2016).

17. Zoe Todd, "Fish Pluralities: Human-Animal Relations and Sites of Engagement in Paulatuuq, Arctic Canada," *Études/Inuit/ Studies* 38, no. 1–2 (2014): 217–38.

18. Dolly Jørgensen has discussed this in the case of organisms that inhabit standing offshore oil structures in "Environmentalists on Both Sides: Enactments in the California Rigs-to-Reefs Debate," in *New Natures: Joining Environmental History with Science and Technology Studies*, ed. Dolly Jørgensen, Finn Arne Jørgensen, and Sara B. Pritchard, 51–68 (Pittsburgh: University of Pittsburgh Press, 2013).

19. Doreen Massey, *For Space* (London: Sage, 2005), 9.

20. Edward Soja, *Thirdspace: Journeys to Los Angeles and Other Real-and-Imagined Places* (London: Wiley-Blackwell, 1996).

21. David J. Bodenhamer, John Corrigan, and Trevor M. Harris, *Deep Maps and Spatial Narratives* (Bloomington: Indiana University Press, 2015); Robert T. Tally, *Spatiality* (London: Routledge, 2013).

22. Henri Lefebvre, *The Production of Space* (London: Wiley-Blackwell, 1991); Nigel Thrift, "Space: The Fundamental Stuff of Geography," in *Key Concepts in Geography*, ed. Sarah L. Holloway, Stephen P. Rice, and Gill Valentine, 95–107 (London: Sage, 2003); Jeremy W. Crampton and Stuart Elden, eds., *Space, Knowledge, and Power: Foucault and Geography* (Aldershot, UK: Ashgate, 2007); Jon Murdoch, *Post-structuralist Geography: A Guide to Relational Space* (London: Sage, 2006).

23. See for instance Bathsheba Demuth, *Floating Coast: An Environmental History of the Bering Strait* (New York: W.W. Norton, 2019); Megan Black, *The Global Interior: Mineral Frontiers and American Power* (Cambridge, MA: Harvard University Press, 2019).

24. William Cronon, *Nature's Metropolis: Chicago and the Great West* (New York: W.W. Norton, 1995).

25. Øyvind Eide, *Media Boundaries and Conceptual Modelling: Between Texts and Maps* (Basingstoke, UK: Palgrave Macmillan, 2015), 23.

26. There is significant scholarship on urban animals and their often strained relationship with urban humans. A good overview of the scholarship is Joshua Specht, "Animal History after Its Triumph: Unexpected Animals, Evolutionary Approaches, and the Animal Lens," *History Compass* 14, no. 7 (2016): 326–36. For more recent work, see the essays in Joanna Dean, Darcy Ingram, and Christabelle Sethna, eds., *Animal Metropolis: Histories of Human-Animal Relations in Urban Canada* (Calgary, AB: University of Calgary Press, 2017); Clemens Wischermann, Aline Steinbrecher, and Philip Howell, eds., *Animal History in the Modern City: Exploring Liminality* (London: Bloomsbury Academic, 2019).

27. Chris Philo and Chris Wilbert, *Animal Spaces, Beastly Places: New Geographies of Human-Animal Relations* (London: Routledge, 2000).

28. Julie Urbanik, *Placing Animals: An Introduction to the Geography of Human-Animal Relations* (Lanham, MD: Rowman & Littlefield, 2012), 183.

29. Urbanik, *Placing Animals*, 185.

30. Timothy Hodgetts and Jamie Lorimer, "Animals' Mobilities," *Progress in Human Geography* 44, no. 1 (2020): 4–26.

31. Derek Gregory and John Urry, eds., *Social Relations and Spatial Structures* (London: MacMillan, 1985), 3.

32. Nicholas Carr, *The Shallows: What the Internet Is Doing to Our Brains* (New York: W. W. Norton, 2010); Sherry Turkle, *Reclaiming Conversation: The Power of Talk in a Digital Age.* (New York: Penguin Books, 2015).

33. Melvin Kranzberg, "Technology and History: 'Kranzberg's Laws,'" *Technology and Culture*, 27, no. 3 (1986): 544–60.

34. Bruno Latour, *Reassembling the Social: An Introduction to Actor-Network Theory* (Oxford: Oxford University Press, 2005), 39.

35. Zeb Tortorici, "Stereoscopic Animals: Spectatorship, Kodiak Bears, and the Keystone *Animal Set*," in *Zoo Studies: A New Humanities*, ed. Tracy McDonald and Daniel Vandersommers, 119–44 (Montreal: McGill-Queens University Press, 2019), 123.

36. Susan R. Schrepfer and Phil Scranton, eds., *Industrializing Organisms: Introducing Evolutionary History* (New York: Routledge, 2004); Edmund Russell, *Greyhound Nation: A Coevolutionary History of England, 1200–1900* (Cambridge: Cambridge University Press, 2018). See also Dolly Jørgensen, "Not by Human Hands: Five Technological Tenets for Environmental History in the Anthropocene," *Environment and History* 20, no. 4 (2014): 479–89.

37. Håkon Stokland, "Field Studies in Absentia: Counting and Monitoring from a Distance as Technologies of Government in Norwegian Wolf Management (1960s–2010s)," *Journal of the History of Biology* 48, no. 1 (2015): 1–36.

38. Jørgensen, *Recovering Lost Species in the Modern Age*.

39. Douglas Ian Campbell and Patrick Michael Whittle, *Resurrecting Extinct Species: Ethics and Authenticity* (Cham, Switzerland: Palgrave Macmillan, 2017), 99–115.

CHAPTER 1: COOPERATIVE CALORIMETRY AND THE INDUSTRIALIZATION OF CATTLE FEEDING

1. Figure 1 shows a different animal, named Steer K, wearing the harness. K's photograph is well documented in a lab book, and his experiment is referred to in published scientific articles. See *Log Book #150, 1911–1915*, 72, PSUA 44, box 2, Animal Nutrition Institute Records, Pennsylvania State University Special Collections (hereafter ANI Records); Donald Cochrane, "The Determination of Ammonia Nitrogen in Steer's Urine," *Journal of Biological Chemistry* 23, no. 1 (1915): 311–16.

2. Based on sensory recollections from Dr. Truman Hershberger, oral history interview by author, August 17, 2016.

3. Rumored in "Science to Aid Stock Raisers," *Chicago Tribune*, February 18, 1900.

4. The connection between temperament and heat production will be featured later in this chapter. "Calm" and "anxious" were terms used by scientists to describe animal demeanor. Researchers have also cited humans exhibiting similar temperament-to-body-heat scenarios. One anecdote by Truman Hershberger described a scenario where young male subjects in the calorimeter demonstrated heightened heat production at the sight of their girlfriends bringing their prepared lunches. He claimed this impacted how they moved forward with experiments. See Brad Olson, "An Oral History of the Armsby Calorimeter at Penn State," February 11, 2014, YouTube video, 15:17, https://www.youtube.com/watch?v=hdxXELdaCDs.

5. Helen Zoe Veit, *Modern Food, Moral Food: Self-Control, Science, and the Rise of Modern American Eating in the Early Twentieth Century* (Chapel Hill: University of North Carolina Press, 2013); Benjamin R. Cohen, *Pure Adulteration: Cheating on Nature in the Age of Manufactured Food* (Chicago: University of Chicago Press, 2019).

6. For some examples, see "The Stock Scare," *Sedalia (MO) Weekly Bazoo*, March 25, 1884; "The Grange," *Willmar (MN) Tribune*, June 27, 1903; "Danger in Rice Hulls," *Fort Wayne (IN) Sentinel*, September 30, 1905.

7. See *Pure Food Legislation: Hearings Before the Committee on Interstate and Foreign Commerce of the House of House of Representatives on Bills H.R. 3109, 12348, 9352, 276 and 4342, Providing against the Adulteration or Misbranding of Foods, Beverages, Drugs, Etc., in the District of Columbia and the Territories, and for Regulating Interstate Traffic in Such Products* (Washington, DC: US Government Printing Office, 1902); United States Federal Trade Commission, *Report of the Federal Trade Commission on Commercial Feeds: March 29, 1921* (Washington, DC: US Government Printing Office, 1921).

8. The approved version enforced pure food for "man and other animals," with detailed feed-related clauses; *Pure Food Legislation*.

9. Winfred Braman, "The Respiration Calorimeter," *Pennsylvania State College Experiment Station Bulletin*, no. 302 (November 1933): 2; U. G. Houck, "The Bureau of

Animal Industry in the United States Department of Agriculture: Its Establishment, Achievements, and Current Activities," 1924, box 1, USDA Bureau of Dairy Industry Records, Special Collections, US National Agricultural Library, Beltsville, MD.

10. Virginia DeJohn Anderson, *Creatures of Empire: How Domestic Animals Transformed Early America* (Oxford: Oxford University Press, 2006); Joshua Specht, *Red Meat Republic: A Hoof-to-Table History of How Beef Changed America* (Princeton, NJ: Princeton University Press, 2019).

11. Specht, *Red Meat Republic*, 239–44.

12. Deborah Valenze, *Milk: A Local and Global History* (New Haven, CT: Yale University Press, 2012); Kendra Smith-Howard, *Pure and Modern Milk: An Environmental History since 1900* (Oxford: Oxford University Press, 2013).

13. See Emily Pawley, "Accounting with the Fields: Chemistry and Value in Nutriment in American Agricultural Improvement, 1835–1860," *Science as Culture* 19, no. 4 (December 1, 2010): 461–82.

14. See Daniel P. Todes, *Pavlov's Physiology Factory: Experiment, Interpretation, Laboratory Enterprise* (Baltimore: Johns Hopkins University Press, 2001); Karen Rader, *Making Mice: Standardizing Animals for American Biomedical Research, 1900–1955* (Princeton, NJ: Princeton University Press, 2004); Tiago Saraiva, *Fascist Pigs: Technoscientific Organisms and the History of Fascism* (Cambridge, MA: MIT Press, 2016).

15. Donna J. Haraway, *Primate Visions: Gender, Race, and Nature in the World of Modern Science* (New York: Routledge, 1989); Gregg Mitman, *The State of Nature: Ecology, Community, and American Social Thought, 1900–1950* (Chicago: University of Chicago Press, 1992); Erika Lorraine Milam, *Creatures of Cain: The Hunt for Human Nature in Cold War America* (Princeton, NJ: Princeton University Press, 2018).

16. Susan Schrepfer and Philip Scranton, *Industrializing Organisms: Introducing Evolutionary History* (New York: Routledge, 2004).

17. See the correspondence between Armsby and Atwater, April 21, 1898–May 8, 1898, box 14, Storrs Agricultural Experiment Station (SAES) Records, Thomas J. Dodd Research Center, University of Connecticut, Storrs. Thank you to Betsy Pittman for her assistance in obtaining these records.

18. Winfred Braman, "The Respiration Calorimeter," *Pennsylvania State College Experiment Station Bulletin*, no. 302 (November 1933): 2.

19. "Peculiar Instrument," *Pittsburgh Press*, December 20, 1901.

20. Susan E. Lederer, *Subjected to Science: Human Experimentation in America before the Second World War* (Baltimore: Johns Hopkins University Press, 1997), 33.

21. "Science to Aid Stock Raisers," *Chicago Tribune*, February 18, 1900.

22. The Brown Dog Affair was a political controversy over vivisection, particularly of a small dog at the University of London, that lasted from 1903 to 1910. See Ben Garlick, "Not All Dogs Go to Heaven, Some Go to Battersea: Sharing Suffering and the "Brown Dog Affair," *Social & Cultural Geography* 16, no. 7 (2015): 798–820; Coral Landsbury, *The Old Brown Dog: Women, Workers, and Vivisection in Edwardian England* (Madison: University of Wisconsin Press, 1985).

23. *Log Book #150, 1911–1915*, 88.

24. The respiration chamber at Möckern (also spelled Mochern) was built in 1879, and was solely an indirect calorimeter. See Frank Liebert, "Tierernahrungsforschung in Leipzig- Möckern (seit Gustav Kuhn) und gegenwartige Forschungsschwerpunkte in der Tierernahrung," *160 Jahre Versuchsstation Möckern*, pamphlet of a presentation given in 2015, originally published in *Feed Magazine*, July–August 2015; Liberty Hyde Bailey, *Cyclopedia of Farm Animals* (New York: Macmillan, 1922), 61.

25. Marion Nestle and Malden Nesheim, *Why Calories Count: From Science to Politics* (Berkeley: University of California Press, 2012), 42.

26. "Energy Values of Red Clover Hay and Maize," *USDA Bureau of Animal Industry*, no. 74 (1905): 25, Armsby wrote that 0.2°C designated a difference of 106 calories.

27. Based on anecdotes from Truman Hershberger, oral history. See Olson, "Oral History of the Armsby Calorimeter." See also Winfred Braman, "The Respiration Calorimeter," *Pennsylvania State College Experiment Station Bulletin*, no. 302 (November 1933): 34.

28. See the list given in *Pennsylvania State College Agricultural Experiment Station Bulletin*, no. 73 (September 1905): 2.

29. Jennifer Karns Alexander, *The Mantra of Efficiency: From Waterwheel to Social Control* (Baltimore: Johns Hopkins University Press, 2008), 79.

30. For example, Henry P. Armsby, "Problems of Animal Nutrition" (paper presented to the Association of American Agricultural Colleges and Experiment Stations meeting, November 1906), Henry P. Armsby Papers, Pennsylvania State University Special Collections, University Park.

31. H. P. Armsby, "The President's Annual Address," *American Society of Animal Nutrition* [1] (1909): 6.

32. The Hatch Act provided federal funds to state land-grant colleges to establish agricultural experiment stations. See Alan I. Marcus, *Agricultural Science and the Quest for Legitimacy: Farmers, Agricultural Colleges and Experimental Stations, 1870–1890* (Ames: Iowa State University Press, 1985).

33. Armsby, "President's Annual Address," printed as H. P. Armsby, "The Food Supply for the Future," *Science* 30, no. 780 (December 10, 1909): 817–23.

34. For example, "Fodder for Cattle, All Grain for Men," *Adams County (PA) Independent*, February 12, 1910; "Animals Eat Our Food," *Inter Ocean* (Chicago), January 9, 1910.

35. See Alan L. Olmstead and Paul W. Rhode, *Creating Abundance: Biological Innovation and American Agricultural Development* (Cambridge: Cambridge University Press, 2008).

36. Henry Prentiss Armsby, "The Maintenance Ration of Cattle," *Pennsylvania State College Agricultural Station Bulletin*, no. 111 (1911).

37. This recalls work in the history of technology related to how animals can be narrated, and have been perceived in the past, as technologies. See Susan Schrepfer and Philip Scranton, eds., *Industrializing Organisms: Introducing Evolutionary History* (New

York: Routledge, 2003); Ann Norton Greene, *Horses at Work: Harnessing Power in Industrial America* (Cambridge, MA: Harvard University Press, 2008).

38. Armsby, "Maintenance Ration of Cattle," 4.

39. For more on the significance of the machine metaphor for modern science, see Anson Rabinbach, *The Human Motor: Energy, Fatigue, and the Origins of Modernity* (Berkeley: University of California Press, 1992).

40. Kendra Smith-Howard speaks further of the significance of cattle viewed as machines and factories in the mid-twentieth century in *Pure and Modern Milk*.

41. Fitzgerald, *Every Farm a Factory*.

42. Armsby, "The Conservation of the Food Supply," *Popular Science Monthly*, November 1911, 499.

43. Alexander, *Mantra of Efficiency*, 81.

44. Just as the header for this section plays off the title of Deborah Kay Fitzgerald, *Every Farm a Factory: The Industrial Ideal in American Agriculture* (New Haven, CT: Yale University Press, 2003).

45. Based on descriptions of where cattle came from for feed experiments, see *Penn State Farmer*, March 1908, 69.

46. Armsby, "Energy Values of Red Clover and Maize Meal."

47. For a current study of this phenomenon, see Jeffrey Rushen et al., "Human Contact and the Effects of Acute Stress on Cows at Milking," *Applied Animal Behaviour Science*, no. 73 (August 1, 2001): 10.

48. The regional specificity of this feeding experiment was important to the researchers and emphasizes earlier theories that feed and hay cultivated in different regions carried different nutritional properties.

49. "Complete Record of Adam's Use of Food Steers," *Log Book #141*, 54, box 2, ANI Records.

50. Armsby, "Maintenance Ration of Cattle," 16.

51. For late twentieth-century examples of this balance between scientific research and animal welfare, see Larry Carbone, *What Animals Want: Expertise and Advocacy in Laboratory Animal Welfare Policy* (Oxford: Oxford University Press, 2004).

52. See J. Thomas Reid, "Energy Values of Feeds—Past, Present, and Future," dedication ceremony of Frank B. Morrison Hall, 1962, 74, Cornell University Agricultural Experiment Station, Ithaca, NY.

53. J. August Fries, Winfred Braman, and Donald Cochrane, "Relative Utilization of Energy in Milk Production and Body Increase of Dairy Cows," *USDA Department Bulletin*, no. 1281 (December 1924): 2.

54. "Science to Aid Stock Raisers," *Chicago Tribune*, February 18, 1900.

55. Testing milking machines at the college was still in its early stages, with later, more comprehensive research on the pros and cons of such machines appearing in publications a few years after the calorimeter was built. See "Test of a Mechanical Cow Milker," *Pennsylvania State College Agricultural Experiment Station Bulletin*, no. 85 (January 1908).

56. *Log Book #150, 1911–1915*, 88–91.

57. Reid, "Energy Values of Feeds," 74.

58. *Log Book #150, 1911–1915*, 93.

59. J. August Fries, Winfred Braman, and Donald Cochrane, "Relative Utilization of Energy in Milk Production and Body Increase of Dairy Cows," *USDA Department Bulletin*, no. 1281 (December 1924): 4.

60. *Penn State Farmer*, April 1909, 68.

61. Agricultural "improvers" were gentlemen farmers acting as lay scientists. For more on the work of "improvers," see Emily Pawley, *The Nature of the Future: Agriculture, Science, and Capitalism in the Antebellum North* (Chicago: University of Chicago Press, 2020).

62. Fitzgerald, *Every Farm a Factory*, 29.

63. *Penn State Farmer*, October 1909, 133.

64. "Developments in the Respiration Calorimeter," *Popular Science Monthly*, May 1904, 187.

65. "A Possible Plan for the Organization of an American Committee (To Be Incorporated) World's Dairy Congress, United States, 1922," Correspondence and Other Records 1907–1922, H. E. Van Norman Papers, Records of the Bureau of Dairy Industry, Record Group 152, National Archives, College Park, MD.

66. Charles E. Rosenberg, "The Adams Act: Politics and the Cause of Scientific Research," *Agricultural History* 38, no. 1 (1964): 10.

67. For example, anonymous editorials, *Hoard's Dairyman*, September 29, 1916, 323; March 9, 1917, 286; December 21, 1917, 769.

68. These are most likely interpretations from the first experiment on timothy hay in 1902. See Henry Armsby and J. August Fries, "The Available Energy of Timothy Hay," *Proceedings from the Society for the Promotion of Agricultural Science* (1902); "Calorimeter Tests," *Allentown (PA) Leader*, April 9, 1902.

69. "Best Place to Fatten a Steer," *Lima (OH) News*, August 7, 1903.

70. "Experiments on Cattle," *New York Times*, April 7, 1902.

71. Armsby, "Maintenance Ration of Cattle," 19.

72. Armsby, "Maintenance Ration of Cattle," 19.

73. To read more about Winterthur's legacy, see Ed Kee, *Delaware Farming* (Mount Pleasant, SC: Arcadia, 2007).

74. See Joanna Radin, *Life on Ice: A History of New Uses for Cold Blood* (Chicago: University of Chicago Press, 2017); Margaret Derry, *Masterminding Nature: The Breeding of Animals, 1750–2010* (Toronto: University of Toronto Press, 2015).

75. Not unlike the services and products developed by entrepreneurs using germ theory in the early twentieth century. See Nancy Tomes, *The Gospel of Germs: Men, Women, and the Microbe in American Life* (Cambridge, MA: Harvard University Press, 1999).

76. "Better Methods of Dairying Illustrated," *De Laval Monthly*, February 1923, 10–11, box 4, USDA Bureau of Dairy Industry Records, Special Collections, National Agricultural Library.

77. "Service Merchandising," *Feedstuffs*, July 1924, 12.

78. *The Purina Cow Book: A-B-C of Milk Making* (N.p.: Ralston Purina Company, 1921), 24.

79. See R. W. Swift, G. P. Barron, K. H. Fisher, C. E. French, E. W. Hartsook, T. V. Hershberger, E. Keck, et al., "The Effect of High versus Low Protein Equicaloric Diets on the Heat Production of Human Subjects," *Journal of Nutrition* 65 (1958): 89–102.

80. See for example T. Yan, M. G. Porter, and C. S. Mayne, "Prediction of Methane Emission from Beef Cattle Using Data Measured in Indirect Open-Circuit Respiration Calorimeters," *Animal* 3, no. 10 (2009): 1455–62; Octavio Alonso Castelán Ortega et al., "Construction and Operation of a Respiration Chamber of the Head-Box Type for Methane Measurement from Cattle," *Animals* 10, no. 2 (February 2020): 227.

CHAPTER 2: PROTOCOLIZING HUMAN-ANIMAL RELATIONSHIPS IN COLONIAL QINGDAO

1. Sebastian Conrad, *German Colonialism: A Short History* (Cambridge: Cambridge University Press, 2012), 62.

2. Asada Shinji, *Doitsu tochika no Chintao. Keizaiteki jiyushugi to shokuminchi shakai chitsujo* (Tokyo: University of Tokyo Press, 2011); Wolfgang Bauer, *Tsingtao 1914 bis 1931: Japanische Herrschaft, wirtschaftliche Entwicklung und die Rückkehr der deutschen Kaufleute* (Munich: Iudicium, 1999); Ran Gyokuji, *Chintao no toshi keisei shi, 1897–1945: Shijo keizai no keisei to tenkai* (Kyoto: Shibunkaku, 2009).

3. Dorothee Brantz, "The Domestication of Empire: Human-Animal Relations at the Intersection of Civilization, Evolution, and Acclimatization," in *A Cultural History of Animals in the Age of Empire*, ed. Kathleen Kete (Oxford: Berg, 2011), 82–85.

4. Paula Young Lee, ed., *Meat, Modernity, and the Rise of the Slaughterhouse* (Durham, NH: University of New Hampshire Press, 2008).

5. Richard Perren, *Taste, Trade and Technology: The Development of the International Meat Industry since 1840* (Aldershot, UK: Ashgate, 2006); Peter Koolmees, "Veterinary Inspection and Food Hygiene in the Twentieth Century," in *Food, Science, Policy and Regulation in the Twentieth Century*, ed. David F. Smith and Jim Phillips (London: Routledge, 2000).

6. Quoted in George Steinmetz, *The Devil's Handwriting: Precoloniality and the German Colonial State in Qingdao, Samoa, and Southwest Africa* (Chicago: University of Chicago Press, 2007), 407.

7. Steinmetz, *Devil's Handwriting*, 408.

8. Ferdinand von Richthofen, *Kiautschou, seine Weltstellung und voraussichtliche Bedeutung* (Berlin: G. Stilke, 1897), 135.

9. Von Richthofen, *Kiautschou*, 136.

10. Von Richthofen, *Kiautschou*, 136.

11. Max Eggebrecht, *Der tierärztliche Anteil an deutscher Kulturarbeit im Schutzgebiet Kiautschou (China)* (Berlin: Tierärztliche Hochschule Berlin, 1923), 20.

12. Manuel Töpfer, *Der Veterinärdienst des Deutschen Reiches in China von 1898 bis 1914* (Giessen, Germany: DVG, 2010), 20.

13. Ernst Rassau, "Fleischbeschau in Deutsch-China," *Berliner Tierärztliche Wochenschrift* 22 (1899): 263.

14. Eggebrecht, *Der tierärztliche Anteil*, 11–12.

15. Töpfer, *Der Veterinärdienst*, 25.

16. Rassau, "Fleischbeschau," 264.

17. Max Eggebrecht, "Die Schlachthofanlage in Tsingtau," *Zeitschrift für Fleisch- und Milchhygiene* 4, no. 18 (January 1908): 109.

18. Eggebrecht, "Die Schlachthofanlage," 111.

19. Eggebrecht, "Die Schlachthofanlage," 112.

20. F. W. Mohr, *Handbuch für das Schutzgebiet Kiautschou* (Tsingtau [Qingdao], China: Walther Schmidt, 1911), 157.

21. Mohr, *Handbuch für das Schutzgebiet*, 156.

22. Mohr, *Handbuch für das Schutzgebiet*, 159.

23. Mohr, *Handbuch für das Schutzgebiet*, 157.

24. Eggebrecht, *Der tierärztliche Anteil*, 14.

25. Mechthild Leutner and Klaus Mühlhahn, *"Musterkolonie Kiautschou": Die Expansion des Deutschen Reiches in China: Deutsch-chinesische Beziehungen 1897 bis 1914; Eine Quellensammlung* (Berlin: Akademie Verlag, 1997), 387.

26. "Geschäfts-Bericht über das sechste Geschäftsjahr 1904," 10, R8127/4144: Schantung Eisenbahn Gesellschaft, 1899–1929, Bd. 1, Bundesarchiv-Lichterfelde, Berlin (hereafter BArch-B).

27. Geschäfts-Bericht über das achte Geschäftsjahr 1906, 7, R8127/4144: Schantung Eisenbahn Gesellschaft 1899–1929, Bd. 1, BArch-B.

28. *Denkschrift betreffend die Entwicklung des Kiautschou-Gebiets von Oktober 1907 bis Oktober 1908* (Berlin: Reichsdruckerei, 1909), 34, https://digital.staatsbibliothek -berlin.de/werkansicht?PPN=PPN770823904&PHYSID=PHYS_0005.

29. Seitō Shubi Gun Minsei Bu, "Seitō tojyū jo," March 1918, 0027, Ref. B07090858200, Japan Center for Asian Historical Records, https://www.jacar.ar chives.go.jp/aj/www/doc/en/before_browse.html (hereafter JACAR).

30. *Denkschrift betreffend die Entwicklung des Kiautschou-Gebiets von Oktober 1904 bis Oktober 1905* (Berlin, 1906), 32.

31. *Denkschrift betreffend die Entwicklung des Kiautschou-Gebiets von Oktober 1907 bis Oktober 1908*, 64.

32. Max Eggebrecht, "Vorschläge zur Durchführung von Zuchtversuchen mit deutschem Vieh," November 25, 1909, RA86: Veterinarwesen in den deutschen Schutzgebieten, Bd.1, 1903, 1910–1914, BArch-B.

33. Max Eggebrecht, "Vorschläge zur Durchführung," n.p.

34. Max Eggebrecht, "Vorschläge zur Durchführung," n.p.

35. [Illegible], "Stellungnahme zu dem Bericht des G.T. vom 25.11.09, Tsingtau," December 20, 1909, RA86, BArch-B.

36. *Denkschrift betreffend die Entwicklung des Kiautschou-Gebiets von Oktober 1908 bis Oktober 1909* (Berlin, 1910), 36.

37. Eggebrecht, *Der tierärztliche Anteil*, 8.

38. Katarzyna J. Cwiertka, *Modern Japanese Cuisine: Food, Power and National Identity* (London: Reaktion Books, 2006), 24–34.

39. Honda Kōsuke, "Bokuchiku fushin no genin oyobi shōrai," *Jitsugyō no Nihon* 4, no. 2 (1901): 14–18.

40. Seitō Gunsei Sho, *Santō no bussan* (Seitō: Seitō Shimpō Sha, 1917), 42.

41. Seitō Gunsei Sho, *Santō no bussan*, 48.

42. Seitō Gunsei Sho, *Santō no bussan*, 50.

43. Seitō Gunsei Sho, *Santō no bussan*, 60.

44. Yamawaki Keikichi, *Nihon teikoku kachiku densenbyō yobō shi: Hōki no hensen*, vol. 3 (Tokyo: Jūeki chōsa jo, 1938), 2.

45. Yamawaki, *Nihon teikoku kachiku densenbyō yobō shi*, 2.

46. Yamawaki, *Nihon teikoku kachiku densenbyō yobō shi*, 32.

47. See, for example, the textbook by Tsuno Keitarō, *Shokuniku keisatsuron* (Tokyo: Tokyo Jūi Shimpō Sha, 1899).

48. Matsubara Kiichirō, "Santō shō oyobi seitō no chiku sangyō," *Chichi to Niku* 6, no. 6 (1915): 45–51; Matsuo Hiroshi, "Kansei seitō tojō no sōkan," *Chichi to Niku* 6, no. 7 (1915): 29–37.

49. March 1915, Ref. A0401029320: Seito tojyū jo ni kansuru kisoku hōkoku no ken, JACAR.

50. Seitō Shubi Gun Minsei Bu, "Seitō tojyū jo," 0030.

51. Yokoi Tokiyoshi, "Goshū niku no yunyū," *Yomiuri Shinbun*, December 24, 1908.

52. "Goshū niku wa dōka," *Asahi Shimbun*, March 16, 1909.

53. Seitō Shubi Gun Minsei Bu, "Seitō tojyū jo," 0039.

54. "Seitō namaushi namaniku yushutsu kumiai seigan sho," 0094, 1921, Ref. B07090848000, JACAR.

55. "Seitō yori namaushi namaniku nihon muke yushutsu no ken," 1922–1924, 180, Ref. B12082462900, JACAR.

56. *Jūeki chōsa jo hōkoku daisanji hōkoku* (Tokyo: Jūeki chōsa jo, 1923), 32.

57. *Jūeki chōsa jo hōkoku daisanji hōkoku.*

58. "Seitō namaushi namaniku yushutsu kumiai seigan sho," 250.

59. Yamawaki, *Nihon teikoku kachiku densenbyō yobō shi*, 29.

60. Yamawaki, *Nihon teikoku kachiku densenbyō yobō shi.*

CHAPTER 3: TECHNOLOGICAL HUMAN-ANIMAL ADAPTATION AMONG THE PASTORALISTS OF INNER MONGOLIA

This research was funded by the European Union's Horizon 2020 research and innovation program under the Marie Skłodowska-Curie grant agreement No. 893394.

1. H. Russell Bernard and Pertii J. Pelto, "Technology and Social Change. Introduction." In *Technology and Social Change*, ed. H. Russell Bernard and Pertii J. Pelto, 1–10 (Prospect Height, IL: Waveland Press, 1987 [1972]), 5.

2. Carole Ferret, "La figure atemporelle du 'nomade des steppes,'" in *La préhistoire des autres: Perspectives archéologiques et anthropologiques*, ed. Nathan Schlanger and Anne-Christine Taylor, 167–182 (Paris: La Découverte, 2012), 167.

3. Alexandra Lavrillier, "Nomadisme et adaptations sédentaires chez les Évenks de Sibérie postsoviétique: 'Jouer' pour vivre avec et sans chamanes" (PhD diss., École Pratique des Hautes Études, 2005), 216–17.

4. Gloria Snively and John Corsiglia, "Indigenous Science: Proven, Practical and Timeless," in *Knowing Home: Braiding Indigenous Science with Western Science, Book 1*, ed. Gloria Snively and Wanosts'a7 Lorna Williams (Victoria, BC: University of Victoria, 2016), chap. 6.

5. Julian Inglis, *Traditional Ecological Knowledge: Concepts and Cases* (Ottawa: International Program on Traditional Ecological Knowledge, International Development Research Centre, 1993), vi.

6. Annamarie Hatcher, Cheryl Bartlett, Albert Marshall, and Murdena Marshall, "Two-Eyed Seeing in the Classroom Environment: Concepts, Approaches, and Challenges," *Canadian Journal of Science, Mathematics and Technology Education* 9, no. 3 (2009): 141–153.

7. The terms are often used as synonyms.

8. Fikret Berkes, *Sacred Ecology* (New York: Routledge, 1999).

9. Consisting of multiple small groups with diverse ethnonyms, the Tungus are scattered throughout Siberia from the Ob to the Okhotsk Sea, in the north of China, and Mongolia. In China they are officially represented by four "ethnic minorities": the Manchu (10,423,303), the Evenki (34,617), the Orochen (9,168), and the Hezhe (5,373) according to *China's* national population census of 2020.

10. Christopher P. Atwood, *Encyclopedia of Mongolia and the Mongol Empire* (New York: Facts on File, 2004), 226.

11. By "Mongolian-speaking Tungus," I refer to the Solon and the Khamnigan peoples, who, through long-term contact with their Mongol neighbors, have adopted the Mongolian language and five-muzzles herding. In the official Chinese classification, the Solon and the Khamnigan are part of the Evenki ethnic minority.

12. Among others, see Douglas Carruthers and Jack Miller Humphrey, *Unknown Mongolia: A Record of Travel and Exploration in North-West Mongolia and Dzungaria* (London: Hutchinson, 1914); Ethel John Lindgren, "North-Western Manchuria and the Reindeer-Tungus," *Geographical Journal* 75, no. 6 (1930): 518–534; Lindgren, "An Example of Culture Contact without Conflict: Reindeer Tungus and Cossacks of North-Western Manchuria," *American Anthropologist*, n.s., 40, no. 4 (1938): 605–621; Ma Hetian 马鹤天, *Nei wai menggu kaocha riji* 内外蒙古考察日记 [Investigation Diary of Inner and Outer Mongolia] (Nanjing: Xin yaxiya xuehui, 1932); Yongfu Mineishan 米内山庸夫 [Yonaiyama Tsuneo], *Menggu fengtu ji* 蒙古风土记 [Record of

Mongolia's Natural Conditions and Social Customs] (Tokyo: Gaizao she, 1938); Sergei Mikhailovich Shirokogoroff, *Social Organization of the Northern Tungus* (New York: Garland, 1979 [1929]); Shirokogoroff, *Psychomental Complex of the Tungus* (London: Kegan Paul, Trench, Trubner, 1935); and Zhang Jiafan张家璠 and Cheng Tingheng 程廷恒, *Hulunbei'er zhilüe* 呼伦贝尔志略 [Notes on Hulun Buir] (1922; repr. Hailar, China: Tianma chuban youxian gongsi, 2003).

13. Shirokogoroff visited numerous Tungus people during his field research in Manchuria and Siberia. Within the framework of this article, the term "Evenki" refers to those living in the modern Hulun Bui area.

14. Shirokogoroff, *Psychomental Complex of the Tungus*, 87.

15. Shirokogoroff, *Psychomental Complex of the Tungus*, 87.

16. David Sneath, "Notions of Rights over Land and the History of Mongolian Pastoralism," *Inner Asia* 3, no. 1 (2001): 48.

17. Hugh Beach and Florian Stammler, "Human-Animals Relations in Pastoralism," *Nomadic Peoples* 10, no. 2 (2006): 12.

18. The Evenki of China possess only domesticated reindeer. There are no wild reindeer in China, in contrast with Siberia, where Evenki reindeer herders hunt wild reindeer.

19. Shirokogoroff, *Social Organization of the Northern Tungus*, 30.

20. Shirokogoroff, *Social Organization of the Northern Tungus*, 36.

21. Beach and Stammler, "Human-Animals Relations in Pastoralism," 7.

22. Shirokogoroff, *Psychomental Complex of the Tungus*, 88.

23. Shirokogoroff, *Psychomental Complex of the Tungus*, 88.

24. Neimenggu zizhiqu bianjizu (NZBZ) 内蒙古自治区编辑组, *Ewenke zu shehui lishi diaocha* 鄂温克族社会历史调查 [Social History Research Investigations on Evenki] (Hohhot, China: Neimenggu renmin chubanshe, 1986), 169.

25. Shirokogoroff, *Psychomental Complex of the Tungus*, 94.

26. NZBZ, *Ewenke zu shehui lishi diaocha*, 275.

27. Natasha Fijn, *Living with Herds: Human-Animal Coexistence in Mongolia* (Cambridge: Cambridge University Press, 2011), 19.

28. Carruthers and Miller, *Unknown Mongolia*, 415.

29. *Pudi* (plural of pood) was a unit of mass broadly used in Russia at the beginning of the twentieth century. It was equivalent to 16.38 kg. Although the USSR abolished the pood in 1924, it was still used among farmers and Russian émigrés who settled in China.

30. Xu Zhanjiang 徐占江 et al., eds., *Zhongguo Buliyate Menggu ren* 中国布里亚特蒙古人 [The Buryat Mongols of China] (Hulunbuir, China: Neimenggu wenhua chubanshe, 2009), 52.

31. Xu et al., *Zhongguo Buliyate Menggu ren*, 52.

32. Anatoly Khazanov, *Nomads and the Outside World* (Cambridge: Cambridge University Press, 1983).

33. Xu et al., *Zhongguo Buliyate Menggu ren*, 52.

34. Shirokogoroff, *Social Organization of the Northern Tungus*, 41.

35. In the present chapter the concept of "modernization" is restricted to the communist era.

36. Cao Fangjun, "Modernization Theory and China's Road to Modernization," *Chinese Studies in History* 43, no. 1 (2009): 8.

37. Caroline Humphrey and David Sneath, *The End of Nomadism? Society, State and the Environment in Inner Asia* (Durham, NC: Duke University Press, 1999), 1.

38. Piers Vitebsky, "Centralized Decentralization: The Ethnography of Remote Reindeer Herders under Perestroika," *Cahiers du Monde Russe et Soviétique* 31, no. 2–3 (1990): 348.

39. Humphrey and Sneath, *End of Nomadism?*, 46–47.

40. The tendency to favor sheep and goat dates back to the 1920s, when the strong foreign market for sheep wool resulted in a rapid growth in the number of sheep and goats; Atwood, *Encyclopedia of Mongolia and the Mongol Empire*, 16.

41. David Sneath, *Changing Inner Mongolia: Pastoral Mongolian Society and the Chinese State* (Oxford: Oxford University Press, 2000), 70.

42. NZBZ, *Ewenke zu shehui lishi diaocha*, 276.

43. Xu Zhanjiang 徐占江 et al., eds., *Zhongguo Mo'ergele Ewenke ren* 中国莫尔格勒鄂温克人 [The Mergel Evenki of China] (Hulunbuir, China: Neimenggu wenhua chubanshe, 2013), 68.

44. Sneath, *Changing Inner Mongolia*, 87.

45. Sneath, *Changing Inner Mongolia*, 87.

46. Sneath, *Changing Inner Mongolia*, 87.

47. Su Jie 苏洁, "Caoyuan chuantong wuzhi wenhua de bianqian: Yi menggubao weili 草原传统物质文化的变迁。以蒙古包为例" [The Transformation of the Grasslands' Traditional Material Culture. A Case Study of the Mongolian Yurt], *Neimenggu Daxue Yishu Xueyuan Xuebao* 3, no. 4 (2010): 55.

48. In some areas, the camel was used, especially for long-distance migration, up to the 1960s.

49. Richard Fraser, "Motorcycles on the Steppe: Skill, Social Change, and New Technologies in Postsocialist Northern Mongolia," *Nomadic Peoples* 22, no. 2 (2018): 39.

50. David S. G. Goodman, "The Campaign to 'Open up the West': National, Provincial-Level and Local Perspectives," *China Quarterly* 178 (2004): 317.

51. Lilian Iselin, "Translocal Practices on the Tibetan Plateau: Motorised Mobility of Pastoralists and Spatial Transformations," *Nomadic Peoples* 18 (2014): 16–17.

52. Humphrey and Sneath, *End of Nomadism?*, 299–300.

53. Alexandra Marois, "D'un habitat mobile à un habitat fixe: Fondements et changements de l'orientation dans l'espace domestique mongol," *Études Mongoles et Sibériennes, Centrasiatiques et Tibétaines* 36–37 (2006): 207–237.

54. Xu et al., *Zhongguo Buliyate Menggu ren*, 74.

55. *Hulunbei'er jijian jiancha ju* 呼伦贝尔纪检监察局, http://hlbe.nmgjjjc.gov.cn/.

56. Humphrey and Sneath, *End of Nomadism?*, 1.

CHAPTER 4: NEGOTIATING HUNTING ETHICS THROUGH TECHNOLOGY

1. Historian Joanna Bourke is a scholar who has devoted most of her interest to the history of killing (of humans). In her works *An Intimate History of Killing: Face-to-Face Killing in 20th Century Warfare* (New York: Basic Books, 1999); and *Deep Violence: Military Violence, War Play, and the Social Life of Weapons* (Berkeley, CA: Counterpoint Press, 2015), she has investigated the gory details of the twentieth-century development of how humans legally kill one another.

2. Bourke, in *Intimate History of Killing*, notes how the mechanization of warfare has transformed the practice during the twentieth century, making fewer people involved in battle capable of killing more efficiently.

3. Ulf Nyrén, "Rätt till jakt: En studie av den svenska jakträtten ca 1600–1789" (PhD diss., Göteborgs Universitet, 2012), 20–21; Matt Cartmill, *A View to a Death in the Morning: Hunting and Nature through History* (Cambridge, MA: Harvard University Press, 1996), 28–51.

4. I have discussed how hunting narratives changed during the nineteenth century in relation to both science and literature, as well as the development of hunting practices in Sweden, in "Changing Narratives of Human-Large Carnivore Encounters," in *Shared Lives of Humans and Animals: Animal Agency in the Global North*, ed. Taina Syrjämaa and Tuomas Räsenen, 163–78 (London: Routledge, 2017).

5. The first explicit questioning of hunting in Sweden appeared in P. A. Säve, "Sista paret ut! (Tankar om behofvet af en lag mellan menniskan och djuren)," *Svenska Jägareförbundets Nya Tidskrift*, no. 2 (1877): 70–86. In the beginning of the twentieth century the geologist, hunter, and explorer Alfred Gabriel Nathorst, published his book *Hafva djuren rättighet att lefva: Några betraktelser* (Stockholm: Wahlström & Widstrand, 1907), which discussed the large-scale killing of animals. Occasionally articles in *Jägaren* also addressed the issue, such as Gustaf Celsing, "En jaktfärd på Spetsbärgen sommaren 1897," *Jägaren*, 1898, 3. *Jägaren* was a twice-yearly magazine, published in a hard-bound edition, containing hunting stories by various authors, as well as notices by the editor, Hugo Samzelius.

6. Thomas Söderqvist, "The Ecologists: From Merry Naturalists to Saviours of the Nation: A Sociologically Informed Narrative Survey of the Ecologization of Sweden 1895–1975" (PhD diss., Göteborgs Universitet, 1986), 26–28.

7. The term "anti-cruelty regime" is borrowed from Per-Anders Svärd, "Problem Animals: A Critical Genealogy of Animal Cruelty and Animal Welfare in Swedish Politics 1844–1944" (PhD diss., Stockholm University, 2015), 253.

8. As the analyzed articles were written in Swedish, the quotes have been translated into English by the author, taking specific care to capture their proper meaning.

9. Karin Dirke, *De Värnlösas Vänner* (Stockholm: Almqvist & Wiksell, 2000), 116–19. See also Roger Bergström et al., "Föreningar och förbund," in *Jaktens historia i Sverige: Vilt—människa—samhälle—kultur*, ed. Kjell Danell, Roger Bergström, Leif Mattsson, and Sverker Sörlin (Stockholm: Liber, 2016), 285–87.

10. Several of the contributors were royal foresters, for example, Erik E:sson Rålamb and Erik von Eckerman.

11. For a discussion on agency and the emergence of posthumanist research methods, see Catherine Adams and Terrie Lynn Thompson, *Researching a Posthuman World: Interviews with Digital Objects* (London: Palgrave Macmillan, 2016).

12. Hugo Samzelius, "Till läsaren," *Jägaren*, 1907, n.p.

13. Hugo Samzelius, "Till läsaren," n.p.

14. Hugo Samzelius, "Till läsaren."

15. Hugo Samzelius, "Till läsaren," *Jägaren*, 1895, 148.

16. See, for example, Berndt Höök, "En björnjakt vid ryska gränsen," *Jägaren* 2 (1896): 21–23; Ernst Tavaststjerna, "På tjäderspel," *Jägaren*, 1897.

17. Gustaf Kolthoff, "En sommar vid Porsangerfjorden," *Jägaren*, 1895, 42–70, relates a trip to Finnmarken where Kolthoff, among other things, shoots birds for collection. Celsing, "En jaktfärd på Spetsbärgen sommaren 1897."

18. Gustaf Kolthoff, "En sommar vid Porsangerfjorden," 43–58; Ad. Patr. Hamilton, "Några räfhistorier," *Jägaren*, 1895, 91. For examples of stories critical about hunting methods or shooting too many animals, see Celsing, "En jaktfärd på Spetsbärgen sommaren"; and Eugen Hemberg, "Skånskt jaktlif," *Jägaren*, 1898, 23.

19. For example, Hugo Samzelius, "Min första orre," *Jägaren*, 1895, 93. For a further discussion on the disciplining of the nineteenth-century hunter, see Dirke: "Changing Narratives of Human-Large Carnivore Encounters."

20. Frida Segerdahl-Nordström, "Min första älgjakt," *Jägaren*, 1898, 123.

21. Hugo Samzelius, "Frida Segerdahl-Nordström," *Jägaren*, 1898, 127.

22. See for example Berndt Höök, "En björnjakt vid ryska gränsen," *Jägaren*, 1896, 23; "Cervus," "Ett jaktminne," *Jägaren*, 1898, 49.

23. See, for example, G. E. R Wasastjerna, "En älgjakt i Ryssland," *Jägaren*, 1895, 72; Henning Nordlund, "En björnjakt i Finland," *Jägaren*, 1895, 100; Frida Segerdahl-Nordström, "Min första älgjakt," *Jägaren*, 1898, 121.

24. See, for example, Samzelius, "Min första orre," 92; Henning Nordlund, "En björnjakt i Finland," *Jägaren*, 1895, 98, 100; [Hugo Samzelius], "Jägarlif i Stockholms och Ålands skärgårdar," *Jägaren*, 1895, 160.

25. Ad. Patr. Hamilton, "Några räfhistorier," *Jägaren*, 1895, 88. Notably he was the same person who above described the use of poison and traps as cruel.

26. Per Larsson, "Skjutvapen," in Danell et al., *Jaktens historia i Sverige*, 67–68, 71–76.

27. [Samzelius], "Jägarlif i Stockholms och Ålands skärgårdar," 160; Carl Kjellin, "Ledsamma bomskott," *Jägaren*, 1895, 179; Ernst Tavaststjerna, "På tjäderspel," *Jägaren*, 1897, 3–4.

28. Gustaf Kolthoff, "En sommar vid Porsangerfjorden," *Jägaren*, 1895, 65.

29. Henning Nordlund, "En senhöstjakt i Norrbotten," *Jägaren*, 1895, 33.

30. Stephanie Rutherford, "A Resounding Success? Howling as a Source of Environmental History," in *Methodological Challenges in Nature-Culture and Environmental*

History Research, ed. Jocelyn Thorpe, Stephanie Rutherford, and L. Anders Sandberg (New York: Routledge, 2017), 43–46. The use of sound in environmental history has been discussed by Peter Coates in "The Strange Stillness of the Past: Toward an Environmental History of Sound and Noise," *Environmental History* 10, no 4 (2005): 636–65.

31. Kolthoff, "En sommar vid Porsangerfjorden," 51.

32. Kolthoff, "En sommar vid Porsangerfjorden," 64.

33. Erik E:sson Rålamb, "Ett jaktminne," *Jägaren*, 1895, 168. Rålamb was a royal forester as well as a painter and sculptor.

34. See, for example, Erik von Eckerman, "Ett efterlängtadt skott," *Jägaren*, 1895, 172; Berndt Höök, "En björnjakt vid ryska gränsen," *Jägaren*, 1896, 23; Valentin Kiörkander, "Lyckad lojakt," *Jägaren*, 1898, 64; Hugo Samzelius, "En holmgång på senhösten," *Jägaren*, 1899, 73; Anna von Eckermann, "På räfjakt," *Jägaren*, 1899, 202.

35. Rålamb, "Ett jaktminne," 166.

36. Henning Nordlund, "En senhöstjakt i Norrbotten," *Jägaren*, 1895, 33. I have also discussed this aspect of hunting stories in Dirke, "Changing Narratives of Human-Large Carnivore Encounters."

37. Nordlund, "En senhöstjakt i Norrbotten," 33; Samzelius, "Min första orre," 95.

38. Frithiof von Sydow, "Skärgårdsminnen och en örnjakt i ungdomsåren" *Jägaren*, 1895, 126.

39. Nordlund, "En senhöstjakt i Norrbotten," 33.

40. Von Eckerman, "Ett efterlängtadt skott," 172.

41. "The birds got four shots sent after them, which resulted in three individuals remaining in the lake." Kolthoff, "En sommar vid Porsangerfjorden," 48.

42. Höök, "En björnjakt vid ryska gränsen," 27.

43. Bourke, *Deep Violence*, 36.

44. Bourke, *Deep Violence*, 17–19, 29.

45. Bourke, *Deep Violence*, 26–29.

46. Nordlund, "En senhöstjakt i Norrbotten," 33. See also Johan Kristian Svanljung, "Ett jaktminne," *Jägaren*, 1899, 152.

47. Rålamb, "Ett jaktminne," 167.

48. Nordlund, "En björnjakt i Finland," 100. In Rålamb, "Ett jaktminne," a wounded fox is killed, after which it is skinned, and seven shots are found in its belly and thigh.

49. [Samzelius], "Jägarlif i Stockholms och Ålands skärgårdar," 159.

50. Nordlund, "En björnjakt i Norrbotten," 184.

51. Wilhelm Liljeborg, "Några erinringar från en vetenskaplig resa," *Jägaren*, 1899, 13. Robert Krig, "En god lohund," *Jägaren*, 1899, 196, is about a hunting dog killing lynx.

52. Bourke, *Deep Violence*, 31–34, 34 (quote).

53. Political scientist Per-Anders Svärd has convincingly argued that the problem of killing animals was constructed in a way that made it possible to eliminate the

inhumane treatment of animals while continuing the overall exploitation of them in society. See Svärd, "Problem Animals: A Critical Genealogy of Animal Cruelty and Animal Welfare in Swedish Politics 1844–1944" (PhD diss., Stockholm University, 2015), 219.

54. On distribution of agency, see Adams and Thompson, *Researching a Posthuman World*, 38–45, 110–12.

55. Thure Månsson, "Samzelius, Johan Axel Hugo," in *Svenska Män och Kvinnor*, vol. 6, ed. Torsten Dahl (Stockholm: Bonniers förlag, 1949), 488–89.

CHAPTER 5: DOGS WITH ANTENNAS AND THE COPRODUCTION OF HUNTING IN A GPS-ENABLED WORLD

1. Åke Nordenram, *Svenska jaktens historia: Från forntid till nutid* (Örkelljunga, Sweden: Bokförlaget Settern, 2001); Carl-Herman Tillhagen, *Allmogejakt i Sverige* (Stockholm: LTs förlag, 1987); James A. Swan, *In Defense of Hunting* (New York: Harper Collins, 1995).

2. Matt Cartmill, *A View to a Death in the Morning: Hunting and Nature through History* (Cambridge, MA: Harvard University Press, 1993).

3. Erica von Essen, "The Impact of Modernization on Hunting Ethics: Emerging Taboos among Contemporary Swedish Hunters," *Human Dimensions of Wildlife* 23, no. 1 (2018): 21–38.

4. Dolly Jørgensen, *Recovering Lost Species in the Modern Age: Histories of Longing and Belonging* (Cambridge, MA: MIT Press, 2019).

5. Irena Knezevic, "Hunting and Environmentalism: Conflict or Misperceptions," *Human Dimensions of Wildlife* 14, no. 1 (2009): 12–20.

6. Julie Urbanik and Connie L. Johnston, eds., *Humans and Animals: A Geography of Coexistence* (Santa Barbara, CA: ABC Clio, 2017), 196.

7. Paul Shepherd, *Man in the Landscape: A Historic View of the Esthetics of Nature* (Athens: University of Georgia Press, 2002), 209.

8. Torolf E. Kroglund, *Jegerhjerte: Jakten på jegerens hemmelighet* (Oslo: Pax, 2015), 10.

9. Lyndel V. Prott, "Hunting as Intangible Heritage: Some Notes on Its Manifestations," *International Journal of Cultural Property* 14, no. 3 (2007): 385–398.

10. Peter S. Alagona, "Introduction: Fifty Years of Wildlife in America," Forum, *Environmental History* 16, no. 3 (2011): 391–397.

11. See, for instance, Daniel Justin Herman, *Hunting and the American Imagination* (Washington, DC.: Smithsonian Institution Press, 2001); Karl Jacoby, *Crimes against Nature: Squatters, Poachers, Thieves, and the Hidden History of American Conservation* (Berkeley: University of California Press, 2001); Tina Loo, *States of Nature: Conserving Canada's Wildlife in the Twentieth Century* (Vancouver: UBC Press, 2006).

12. Christer Lindberg, *Älgjakt: 9 sätt att jaga älg* (Örkelljunga, Sweden: Bokförlaget Settern, 2012).

13. In British English the Eurasian moose is called elk.

14. See Edmund Russell, *Greyhound Nation: A Coevolutionary History of England, 1200–1900* (Cambridge: Cambridge University Press, 2018).

15. Etienne Benson, *Wired Wilderness: Technologies of Tracking and the Making of Modern Wildlife* (Baltimore: Johns Hopkins University Press, 2010).

16. John Staudenmaier, *Technology's Storytellers: Reweaving the Human Fabric* (Cambridge, MA: MIT Press, 1989).

17. Paul E. Ceruzzi, *GPS* (Cambridge, MA: MIT Press, 2018), 7.

18. Bill Clinton, "Statement by the President Regarding the United States' Decision to Stop Degrading Global Positioning System Accuracy," Office of the Press Secretary, The White House, May 1, 2000, https://clintonwhitehouse4.archives.gov/WH/EOP/OSTP/html/0053_2.html.

19. Benson, *Wired Wilderness*.

20. Skogforsk, "Sverige har världens tätaste älgstam," 2022, https://www.skogforsk.se/kunskap/kunskapsbanken/2022/sverige-har-varldens-tataste-algstam/

21. Länsstyrelsen, "Statistik för älgdata," 2019, https://algdata-apps.lansstyrelsen.se/algdata-apps-stat.

22. "Björnjakt: Den svarte på Torpaflen," *Svensk Jakt*, August 2012, 15.

23. Naturvårdsverket, "Varg, population Skandinavien," accessed November 14, 2022, https://www.naturvardsverket.se/Sa-mar-miljon/Vaxter-och-djur/Rovdjur/Fakta-om-varg/.

24. Leena Koski and Pia Bäcklund, "Whose Agency? Humans and Dogs in Training," in *Shared Lives of Humans and Animals: Animal Agency in the Global North*, 11–23, ed. Tuomas Räsänen and Taina Syrjämaa (London: Routledge, 2017).

25. "Älgjakt: Magisk måndag i Finnfara," *Svensk jakt*, October 2012, 24.

26. Kjell Danell et al., eds., *Jaktens historia i Sverige: Vilt—människa—samhälle—kultur* (Stockholm: Liber, 2016), 119.

CHAPTER 6: WOLF RESEARCH AS MULTISPECIES KNOWLEDGE PRODUCTION IN FINLAND

This paper is part of the project Disappeared, Endangered, and Newly Arrived Species (341118) funded by the Academy of Finland. I would like to thank the reviewers for their valuable comments.

1. John Berger, "Why Look at Animals?," in *Animals and Society: Critical Concepts in the Social Sciences*, ed. Rhoda Wilkie and David Inglis (London: Routledge, 2007), 72.

2. Donna Haraway, *When Species Meet* (Minneapolis: University of Minnesota Press, 2008), 25, 26. Bruno Latour has also used the term "natureculture": Bruno Latour, *Politics of Nature: How to Bring the Sciences into Democracy*, trans. Catherine Porter (Cambridge, MA: Harvard University Press, 2004).

3. Bruno Latour, *The Pasteurization of France*, trans. Alan Sheridan (Cambridge, MA: Harvard University Press, 1993); Michel Callon, "Some Elements of a Sociology

of Translation: Domestication of the Scallops and the Fishermen of St Brieuc Bay," in *Power, Action and Belief: A New Sociology of Knowledge?*, ed. John Law, 196–233 (London: Routledge & Kegan Paul, 1986).

4. Lynda Birke, Mette Bryld, and Nina Lykke, "Animal Performances: An Exploration of Intersections between Feminist Science Studies and Studies of Human/Animal Relationships," *Feminist Theory* 5, no. 2 (2004): 167–83.

5. In the 1960s biologist Erkki Pulliainen studied the wolf expansion taking place at the time. The research was first conducted by the Game Management Foundation (Riistanhoito-Säätiö) and later, from the 1970s on, by the Finnish Game Research Institute (Riista- ja kalatalouden tutkimuslaitos). Heta Lähdesmäki, *Susien paikat: Ihminen ja susi 1900-luvun Suomessa*, Nykykulttuurin tutkimuskeskuksen julkaisuja 127 (Jyväskylä, Finland: University of Jyväskylä, 2020); originally a PhD dissertation at the University of Turku.

6. Luonnonvarainneuvosto, *Suomen suurpetokannat ja niiden hoito* (Helsinki: Maa- ja metsätalousministeriö, 1998), 12; Suurpetotyöryhmä, *Suomen maasuurpetokannat ja niiden hoito: Suurpetotyöryhmän raportti* (Helsinki: Maa- ja metsätalousministeriö, 1996), 12. Observations by volunteers and field staff are still perceived as a good method for gathering information on wolves in Finland. "Method of Estimating Finnish Wolf Population Size Has Been Evaluated," Natural Resources Institute Finland (Luke), October 21, 2016, https://www.luke.fi/en/news/method-of-estimating-finnish-wolf-population-size-has-been-evaluated/.

7. Luonnonvarainneuvosto, *Suomen suurpetokannat ja niiden hoito*, 15–16. All translations from the Finnish in this chapter are by the author.

8. Ilpo Kojola et al., "Predation on European Wild Forest Reindeer (*Rangifer tarandus*) by Wolves (*Canis lupus*) in Finland," *Journal of Zoology* 263, no. 3 (2004): 232.

9. See for instance, Mika Parkkonen, "Susien liikkeitä seurataan tutkalla Kuhmossa," *Helsingin Sanomat*, July 23, 1998; Ilpo Kojola, "Susille radiopantoja Kainuussa," *Metsästäjä*, May 1998, 30.

10. Ilpo Kojola et al., "Dispersal in an Expanding Wolf Population in Finland," *Journal of Mammalogy* 87, no. 2 (2006): 282. During 1998–2000 eight wolves from three packs (three alpha couples and two pups) were radio-collared in the Kainuu region. Kojola et al., "Predation on European Wild Forest Reindeer," 230.

11. David Mech and Luigi Boitani, "Preface," in *Wolves: Behavior, Ecology and Conservation*, ed. David Mech and Luigi Boitani (Chicago: University of Chicago Press, 2006), xi–xii.

12. Kojola et al., "Dispersal in an Expanding Wolf Population," 282; Kojola et al., "Predation on European Wild Forest Reindeer," 230. About radiotelemetry in general, see Mark R. Fuller and Todd K. Fuller, "Radio-Telemetry Equipment and Applications for Carnivores," in *Carnivore Ecology and Conservation: A Handbook of Techniques*, ed. Luigi Boitani and Roger A. Powell, 152–168 (New York: Oxford University Press, 2012).

13. Mech and Boitani, "Preface," xi; Jon T. Coleman, *Vicious: Wolves and Men in America* (New Haven, CT: Yale University, 2004), 81–82.

14. Etienne Benson, *Wired Wilderness: Technologies of Tracking and the Making of Modern Wildlife* (Baltimore: Johns Hopkins University Press, 2010), 2, 5–51, 56; Etienne Benson, "The Wired Wilderness: Electronic Surveillance and Environmental Values in Wildlife Biology" (PhD diss., Massachusetts Institute of Technology, 2008).

15. Benson, *Wired Wilderness*, 2.

16. Fuller and Fuller, "Radio-Telemetry Equipment," 162.

17. See European Union, Council Directive 92/43/EEC of May 21, 1992, on the conservation of natural habitats and of wild fauna and flora.

18. Riista ja kalatalouden tutkimuslaitos, *RKT Vuosikertomus 1998* (Helsinki: Riista ja kalatalouden tutkimuslaitos 1999), 3; Riista ja kalatalouden tutkimuslaitos, *RKT Toimintakertomus 1999* (Helsinki: Riista ja kalatalouden tutkimuslaitos 2000), 18; Heta Lähdesmäki, "Hallittu ja hallitsematon susi: Valta ihmisen ja suden välisessä suhteessa poronhoitoalueen ulkopuolisessa Suomessa 1990-luvun lopulla" (master's thesis, University of Turku, 2011), 51, 52, 55.

19. Kojola, "Susille radiopantoja Kainuussa," 30.

20. Fuller and Fuller, "Radio-Telemetry Equipment," 153.

21. Berger, "Why Look at Animals?," 72.

22. Lori Gruen and Kari Weil, "Teaching Difference: Sex, Gender, Species," in *Teaching the Animal: Human–Animal Studies across the Disciplines*, ed. Margo DeMello (New York: Lantern Books, 2010), 128.

23. Coleman, *Vicious*, 81–82.

24. Kojola et al., "Dispersal in an Expanding Wolf Population," 282; Kojola et al., "Predation on European Wild Forest Reindeer," 230. On the media coverage, see, for instance, Parkkonen, "Susien liikkeitä seurataan"; Finnish News Agency (STT), "Ilomantsin ja Kuhmon susille asetetaan radiolähetinpantoja," *Helsingin Sanomat*, January 1, 1999; Erkki A. Kauhanen, "Metsässä kello 10," *Helsingin Sanomat*, January 27, 2001.

25. Alexander Pschera, *Animal Internet: Nature and the Digital Revolution*, trans. Elisabeth Lauffer (New York: New Vessel Press, 2014), 94.

26. Birke, Bryld, and Lykke, "Animal Performances," 169; Judith Butler, *Gender Trouble: Feminism and the Subversion of Identity* (London: Routledge, 1990).

27. Karen Barad, "Posthumanist Performativity: Toward an Understanding of How Matter Comes to Matter," *Signs: Journal of Women in Culture & Society* 28, no. 3 (2003): 803.

28. Barad, "Posthumanist Performativity," 829.

29. Birke, Bryld, and Lykke, "Animal Performances," 167, 172.

30. Kojola et al., "Dispersal in an Expanding Wolf Population," 282; and Kojola et al., "Predation on European Wild Forest Reindeer," 230–31. See also Erkki Pulliainen and Lassi Rautiainen, *Suurpetomme: Karhu, susi, ahma, ilves* (Kotka, Finland: Articmedia, 1999), 174.

31. Pschera, *Animal Internet*, 6.

32. Birke, Bryld, and Lykke, "Animal Performances," 175.

33. Jonna Karjalainen, "Seppo on alistanut suden 300 kertaa," *Yle News*, April 22, 2017.

34. Karjalainen, "Seppo on alistanut suden."

35. Kojola et al., "Predation on European Wild Forest Reindeer," 231.

36. Salla Kaartinen et al., "Finnish Wolves Avoid Roads and Settlements," *Annales Zoologici Fennici* 42, no. 5 (2005): 525.

37. Parkkonen, "Susien liikkeitä seurataan," 5.

38. Tapio Mainio, "Yksinäisen gps-suden kosiomatka," *Helsingin Sanomat*, April 6, 2004.

39. Pulliainen and Rautiainen, *Suurpetomme*, 174. See also Pschera, *Animal Internet*, 6.

40. Mainio, "Yksinäisen gps-suden kosiomatka."

41. Kojola et al., "Dispersal in an Expanding Wolf Population," 282.

42. Fuller and Fuller, "Radio-Telemetry Equipment," 164.

43. Pulliainen and Rautiainen, *Suurpetomme*, 174; Mech and Boitani, "Preface," xi. See also Pschera, *Animal Internet*, 6, 7.

44. Kojola et al., "Dispersal in an Expanding Wolf Population," 282.

45. Kojola et al., "Dispersal in an Expanding Wolf Population," 283.

46. Kojola et al., "Dispersal in an Expanding Wolf Population," 282.

47. STT, "Sudet hakeutuvat läntisemmille asuma-alueille," *Helsingin Sanomat*, May 3, 2002.

48. Mainio, "Yksinäisen gps-suden kosiomatka"; Jukka Kykkänen, "Etelä-Pohjanmaan sudet peräisin P-Karjalasta," *Maaseudun Tulevaisuus*, November 26, 2003.

49. Lähdesmäki, *Susien paikat*, 90.

50. Mainio, "Yksinäisen gps-suden kosiomatka."

51. Kojola et al., "Predation on European Wild Forest Reindeer," 231.

52. Kaartinen et al., "Finnish Wolves Avoid Roads," 523.

53. Karen Barad, *Meeting the Universe Halfway: Quantum Physics and the Entanglement of Matter and Meaning* (Durham, NC: Duke University Press, 2006), 21, 33, 118, 178. See also Birke, Bryld, and Lykke, "Animal Performances," 178.

54. Barad, "Posthumanist Performativity," 828.

55. Barad, "Posthumanist Performativity," 817. See also 810 and 816; and Barad, *Meeting the Universe*, 202.

56. Fuller and Fuller, "Radio-Telemetry Equipment," 163; Parkkonen, "Susien liikkeitä seurataan."

57. Parkkonen, "Susien liikkeitä seurataan,"

58. Kojola et al., "Dispersal in an Expanding Wolf Population," 285; Fuller and Fuller, "Radio-Telemetry Equipment," 155.

59. STT, "Ilomantsin ja Kuhmon susille."

60. Fuller and Fuller, "Radio-Telemetry Equipment," 156.

61. Parkkonen, "Susien liikkeitä seurataan," 5.

62. Mainio, "Yksinäisen gps-suden kosiomatka."

63. Fuller and Fuller, "Radio-Telemetry Equipment," 157–59.

64. Parkkonen, "Susien liikkeitä seurataan," 5.

65. Benson, *Wired Wilderness*, 80–81.

66. Pschera, *Animal Internet*, 102.

67. Terry J. Kreeger, "The Internal Wolf: Physiology, Pathology, and Pharmacology," in Mech and Boitani, *Wolves*, 216.

68. Kojola et al., "Dispersal in an Expanding Wolf Population," 282; Kojola et al., "Predation on European Wild Forest Reindeer," 230–31.

69. Kaartinen et al., "Finnish Wolves Avoid Roads," 525.

70. Haraway, *When Species Meet*, 23–25.

71. Fuller and Fuller, "Radio-Telemetry Equipment," 153.

72. Pschera, *Animal Internet*, 103.

73. Kojola et al., "Predation on European Wild Forest Reindeer"; Ilpo Kojola et al., "Interactions between Wolves *Canis lupus* and Dogs *Canis familiaris* in Finland," *Wildlife Biology* 10, no. 2 (2004): 101–6. See also Ilpo Kojola and Seppo Ronkainen, "Susia radioseurannassa Kainuussa," *Metsästäjä*, June 1998): 12–14; Maija Orava, "Tuhat karhua ja sata sutta," *Helsingin Sanomat*, April 18, 1999; Parkkonen, "Susien liikkeitä seurataan." Similar results of individual behavior have been found in other countries when studying other large carnivores. See John D. C. Linnell et al., "Large Carnivores That Kill Livestock: Do 'Problem Individuals' Really Exist?" *Wildlife Society Bulletin* 27, no. 3 (1999): 701.

74. STT, "Ilomantsin ja Kuhmon susille."

75. Kaartinen et al., "Finnish Wolves Avoid Roads," 527, 529.

76. Kojola et al., "Dispersal in an Expanding Wolf Population," 284.

77. As Coleman writes, radio-collared wolves have provided surprising information on dispersal in North America as well. For instance, the great distances wolves can disperse has surprised researchers there. Coleman, *Vicious*, 23.

78. Birke, Bryld, and Lykke, "Animal Performances," 167, 172.

79. Pschera, *Animal Internet*, 100.

80. Pschera, *Animal Internet*, 9.

81. Kojola et al., "Dispersal in an Expanding Wolf Population," 282.

82. Tina Loo, "The Bow Valley and 'People' without a History," *NiCHE*, January 10, 2018, https://niche-canada.org/2018/01/10/the-bow-valley-and-people-without-a-history/.

83. Colleen Campbell and Tina Loo, "Making Tracks: A Grizzly and Entangled History," in *Traces of the Animal Past: Methodological Challenges in Animal History*, ed. Jennifer Bonnell and Sean Kheraj, 235–68 (Calgary, AB: University of Calgary Press, 2022).

84. Parkkonen, "Susien liikkeitä seurataan," 5; Anne Helakoski and Anne Leppänen, "Pyhäjoelle kasvaa susilauma: Pentuja odottava pantasusi Milla kumppaneineen huolettaa koiraharrastajia," *Kaleva*, May 3, 2002; Karjalainen, "Seppo on alistanut suden"; Pulliainen and Rautiainen, *Suurpetomme*, 174; Marja Valtavaara, "Kuhmon

pantasusien kaatoluvat närkästyttävät," *Helsingin Sanomat*, November 1, 2003; Mainio, "Yksinäisen gps-suden kosiomatka." Some of these are common Finnish names, such as Maija, and others are rare, like Ukri and Ugri, which have a Fenno-Ugrian ring to them, as well as Russian names such as Igor. From the names mentioned here, only Noppe is used for dogs, not humans.

85. See for instance Benson, *Wired Wilderness*, 59, 80.

86. Jeffrey Moussaieff Masson and Susan McCarthy, *When Elephants Weep: The Emotional Lives of Animals* (London: Vintage, 1995), 47–48.

87. Jukka Bisi and Sami Kurki, *Susipuhetta Suomessa: Maakunnalliset ja kansalliset odotukset ja tavoitteet susikannan hoidossa* (Seinäjoki: Helsingin yliopisto, maaseudun tutkimus- ja koulutuskeskus, 2005), 141.

88. Minna Polkki, "Ministeriö valmistelee susien suojelun hoitosuunnitelmaa," *Helsingin Sanomat*, February 4, 2004.

89. Pulliainen and Rautiainen, *Suurpetomme*, 174.

90. Olli Puustinen, "Susiasenteissa olisi kohentamista," *Helsingin Sanomat*, editorial, February 3, 2001.

91. Ritva Liikkanen, "Kainuussa nousee susiviha ja laki uhataan ottaa omiin käsiin," *Helsingin Sanomat*, April 2, 2000.

92. Frans de Waal, *Primates and Philosophers: How Morality Evolved* (Princeton, NJ: Princeton University Press, 2006), 65.

93. Pschera, *Animal Internet*, 101.

94. Tapio Mainio, "Petoyhdysmiehet lakkoilevat Sotkamossa," *Helsingin Sanomat*, March 20, 2004. See also Kaartinen et al., "Finnish Wolves Avoid Roads," 524.

95. Kojola et al., "Dispersal in an Expanding Wolf Population," 284.

96. STT, "Pantasusi Milla löytyi tapettuna Pyhäjoelta," *Maaseudun Tulevaisuus*, June 23, 2003.

97. STT, "Hirvi tappoi pantasuden Sonkajärvellä," *Helsingin Sanomat*, December 10, 2004.

98. Fuller and Fuller, "Radio-Telemetry Equipment," 167.

99. See for instance STT, "Susien radioseuranta jatkuu," *Helsingin Sanomat*, February 28, 2001.

100. Bisi and Kurki, *Susipuhetta Suomessa*, 61, 66, 76, 82, 108, 109, 121, 128, 140–41.

101. Liikkanen, "Kainuussa nousee susiviha."

102. Besides Ugri, one other wolf from his pack was killed illegally in Kainuu, and three of his pack members were killed legally in Suomussalmi's reindeer herding area. STT, "Yliajettu oli pantasusi Ugri," *Helsingin Sanomat*, March 5, 2001; Matti Huuskonen, "Salakaatajat paloittelivat tutkimussuden Kuhmossa," *Helsingin Sanomat*, March 6, 2001; Matti Huuskonen, "Tutkimussudet kuumentavat tunteita Kainuussa," *Helsingin Sanomat*, March 7, 2001.

103. Huuskonen, "Tutkimussudet kuumentavat tunteita."

104. Lähdesmäki, "Hallittu ja hallitsematon," 91–96; Lähdesmäki, "Susien paikat," 68–70; Bisi, and Kurki, *Susipuhetta Suomessa*, 7, 61, 70, 75, 78, 107, 136, 140.

105. "Poliisi yritti ampua Nopen puolison," *Helsingin Sanomat*, April 6, 2004.

106. Pschera, *Animal Internet*, 100–101.

107. Pschera, *Animal Internet*, 103–4.

108. Coleman, *Vicious*, 82.

109. See also Coleman, *Vicious*, 80.

110. Stephanie Rutherford, *Villain, Vermin, Icon, Kin: Wolves and the Making of Canada* (Montreal: McGill-Queen's University Press, 2022); Karen R. Jones, *Wolf Mountain: A History of Wolves along the Great Divide* (Calgary, AB: University of Calgary Press, 2002).

111. Reijo Vesterinen, "Radiopanta kertoo suden kulkureitin," *Maaseudun Tulevaisuus*, October 6, 2004. No wolves were GPS-collared in Finland during the winters of 2020 and 2021. Technical problems, challenges in getting permissions, and questions related to work safety were named as reasons behind the decision not to collar wolves. Researchers regarded DNA analyses as a more valuable method of studying wolves. Still, hunters and game officials have been in favor of collaring because of the fact that the data it offers can be used to prevent hunting dogs from being attacked by wolves. "Luonnonvarakeskus kehittää susien merkitsemiseen liittyviä toimintamalleja ja tietopalveluita—pannoituksissa välivuosi," Natural Resources Institute Finland (Luke), February 11, 2020, https://www.luke.fi/uutinen/susien-pannoituksissa-valivu osi/; "Susia ei pannoiteta kevättalvella 2021—kanta-arvion tietopohja säilyy vahvana," Natural Resources Institute Finland (Luke), November 16, 2020, https://www.luke .fi/uutinen/susia-ei-pannoiteta-kevattalvella-2021/; "Käyttäjiltä palautetta: Puutteista huolimatta susien pantapalvelua pidetään tärkeänä," Natural Resources Institute Finland (Luke), March 25, 2020, https://www.luke.fi/uutinen/susia-ei-pannoiteta-keva ttalvella-2021/.

CHAPTER 7: TRANSFORMING THE HUMAN-EAGLE RELATIONSHIP THROUGH CONSERVATION TECHNOLOGIES

1. Tuomas Räsänen, "Does a Dead Wild Animal Have Agency? The White-Tailed Eagle as a Catalyst for an Ideational Revolution in Finland," in *Shared Lives of Humans and Animals: Animal Agency in the Global North*, ed. Tuomas Räsänen and Taina Syrjämaa, 93–104 (London: Routledge, 2017).

2. As for the former, see, for example, Björn Helander, ed., *Sea Eagle 2000*, proceedings from the International Sea Eagle Conference in Björkö, Sweden, September 13–17, 2000 (Stockholm: Swedish Society for Nature Conservation, 2003); Torsten Stjernberg, ed., *Projekt havsörn i Finland och Sverige: Förhandlingar från ett havsörnsymposium 8–9.1.1979 på Tvärminne zoologiska station, Finland* (Helsinki: Jord- och skogsbruksministeriet, 1981); Torsten Stjernberg, "Havsörnsskydd med avstamp i finsk medeltid," *Luonnontieteellisen keskusmuseon vuosikirja* 1995 (Helsinki: Luonnontieteellinen

keskusmuseo, 1995); Ismo Nuuja and Kalle Roukolainen, eds., *Merikotkien puolesta: WWF:n merikotkatyöryhmän vuosikymmenten taival* (Helsinki: WWF Suomi, 2017).

3. See, for example, Timothy J. Farnham, *Saving Nature's Legacy: Origins of the Idea of Biodiversity* (New Haven, CT: Yale University Press, 2007); Roderick Frazier Nash, *The Rights of Nature: A History of Environmental Ethics* (Madison: University of Wisconsin Press, 1989).

4. David Conradson, "Geographies of Care: Spaces, Practices, Experiences," *Social & Cultural Geography* 4 (2003): 451; Alex Franklin and Nora Schuurman, "Aging Animal Bodies: Horse Retirement Yards as Relational Spaces of Liminality, Dwelling and Negotiation," *Social & Cultural Geography* 20 (2019): 918–37; Christine Milligan and Janine Wiles, "Landscape of Care," *Progress in Human Geography* 34 (2010): 736–54. María Puig de la Bellacasa, in particular, emphasizes work as one crucial dimension in care, along with affection and ethics/politics; Puig de la Bellacasa, *Matters of Care: Speculative Ethics in More than Human Worlds* (Minneapolis: University of Minnesota Press, 2017), 1–6.

5. Nash, *Rights of Nature*, 49; Dolly Jørgensen, "Artifacts and Habitats," in *The Routledge Companion to Environmental Humanities*, ed. Ursula K. Heise, Jon Christensen, and Michelle Niemann (London: Routledge 2017), 141.

6. Sometimes these activities were quite intensive. In her history of wildlife conservation in Canada, Tina Loo describes how game managers in the early twentieth century supported beaver and muskrat populations, which were valued for their fur, by planting wild rice, digging ditches, and building artificial dams. Loo, *States of Nature: Conserving Canada's Wildlife in the Twentieth Century* (Vancouver: UBC Press, 2006), 110, 117, 128.

7. World Wildlife Fund, *World Wildlife: Yearbook 1970–71*, ed. Peter Jackson (Morges, Switzerland: World Wildlife Fund, 1971), 31–175; Alexis Schwarzenbach, *Saving the World's Wildlife: WWF—The First 50 Years* (London: Profile Books, 2011).

8. Gill Aitken, "Conservation and Individual Worth," *Environmental Values* 6 (1997): 439.

9. See for example, World Wildlife Fund, "Arabian Oryx: Preventing a Second Extinction," accessed September 9, 2023, http://wwf.panda.org/wwf_news/?1938/Arabian-oryx-preventing-a-second-extinction. For rescue and reintroduction in animal conservation, see Anthony R. E. Sinclair, John M. Fryxell, and Graeme Gaughley, *Wildlife Ecology, Conservation and Management* (Chichester, UK: Wiley Blackwell, 2014), 323–24.

10. Bruno Latour, *Reassembling the Social: An Introduction to Actor-Network-Theory* (Oxford: Oxford University Press, 2005), 39.

11. Teuvo Suominen, *Lintujemme katoava aateli* (Porvoo, Finland: WSOY, 1967), 68; Mari Pohja-Mykrä, Timo Vuorisalo, and Sakari Mykrä, "Organized Persecution of Birds of Prey in Finland: Historical and Population Biological Perspectives," *Ornis Fennica* 89 (2012): 11.

12. Heikki Lehikoinen, *Tuo hiisi hirviäsi: Metsästyksen kulttuurihistoriaa Suomessa* (Helsinki: Teos, 2007), 57.

13. Björn Helander, *Havsörnen i Sverige* (Stockholm: Svenska Naturskyddsföreningen, 1975), 5, 16; Suominen, *Lintujemme katoava aateli*, 68–70.

14. For example, at least ten adult eagles were found dead during the severe winter of 1965–1966. This was a serious blow to a population of less than one hundred individuals. Torsten Stjernberg, "Havsörnen i Finland: Rekommendation för skydd av havsörnen boplatser," manuscript, March 10, 1977, Memorandums and Reports, WWF Finland, Working Group on the White-Tailed Eagle (hereafter Working Group).

15. See, for example, Helander, *Havsörnen i Sverige*, 18; Björn Helander, zoologist, Swedish Museum of Natural History, personal communication, November 25, 2015; Torsten Stjernberg, Juhani Koivusaari, and Ismo Nuuja, "Suomen merikotkakannan kehitys ja pesintätulos 1970–89," *Lintumies* 25 (1990): 68–71; Record from the meeting of WWF Finland Working Group on the White-Tailed Eagle (hereafter Meeting record), February 27, 1973, Records 1972–1986, Working Group; Stjernberg, "Havsörnen i Finland."

16. Sören Jensen, personal communication, April 21, 2015; Jensen, "Report of a New Chemical Hazard," *New Scientist* 32 (1966): 612; Jensen, "The PCB Story," *Ambio* 1 (1972): 123–31; "Havsörn hjälpte Jensen," *Miljöaktuellt*, December 12, 2006.

17. Jensen, personal communication; Sören Jensen, A. G. Johnels, Mats Olsson, and Gunnar Otterlind, "DDT and PCB in Marine Animals from Swedish Waters," *Nature* 224 (1969): 247–50; Tuomas Räsänen, "Itämeren ympäristökriisin ja uuden merisuhteen synty Suomessa 1960-luvulta 1970-luvun puoliväliin" (PhD diss., University of Turku, Finland, 2015), 62–73, 90–94.

18. WWF, *World Wildlife: Yearbook 1970–71*, 31–175; Schwarzenbach, *Saving the World's Wildlife*.

19. Helander, personal communication; Heidi Björklund, "Torsten Stjernberg—Monipuolinen eläintieteilijä," *Luonnon Tutkija* 118 (2014): 108.

20. Farnham, *Saving Nature's Legacy*.

21. Helander, *Havsörnen i Sverige*, 27–28; Ismo Nuuja, a former protection coordinator of the WWF Finland Working Group on the White-Tailed Eagle, personal communication, November 2, 2017.

22. See for example, Björn Helander, "Örndata: Projekt HAVSÖRN under 1971 och 1972," *Sveriges Natur* 63 (1972): 290; Meeting record, December 19, 1972; Meeting record, January 14, 1973; Meeting record, May 5, 1976, all in Records 1972–1986, Working Group; Stjernberg, "Havsörnsskydd med avstamp i finsk medeltid," 44.

23. I was informed about the problems of feeding California condors by Jeff Miller, Conservation Advocate at the Center for Biological Diversity, email, April 19, 2013.

24. Noah Cincinnati, "Too Sullen for Survival: Historicizing Gorilla Extinction, 1900–1930," in *The Historical Animal*, ed. Susan Nance, 166–83 (Syracuse, NY: Syracuse University Press, 2015).

25. See, for example, Will Steffen et al., "The Trajectory of the Anthropocene: The Great Acceleration," *Anthropocene Review* 2 (2015): 81–98.

26. Pauli Kleemola, "Use of the Water Courses for Recreational Purposes," *Aqua Fennica* 1 (1971): 62–63.

27. Report by Juhani Koivusaari on the upcoming nest inventories in the Kvarken area in 1984, August 2, 1983, n.p., Nesting Reports and Various Documents, Working Group. Counterintuitive as this quote appears, felling the trees and destroying nests were hoped to encourage the eagles to move to more remote and thus safer areas to nest.

28. See, for example, Helander, *Havsörnen i Sverige*, 24–26; Meeting record, August 23, 1976, Records 1972–1986, Working Group; Torsten Stjernberg, "Havsörnen i Finland"; *The White-Tailed Eagle Protection Strategy*, March 17, 1982, Memorandums and Reports, Working Group.

29. See, for example, *Sea Eagle—Conservation Programme for Northern Europe and Greenland*, undated report, IUCN/WWF project, doc. no. 1972/1973; Meeting record, January 20, 1973, ; Meeting record, September 11, 1982; Meeting record, May 6, 1982, all in Records 1972–1986, Working Group.

30. Meeting record, October 13, 1977, Records 1972–1986, Working Group. See also Henrik Wallgren, memorandum, n.d. [1985], Memorandums and Reports, Working Group.

31. Meeting record, October 14, 1974, Records 1972–1986, Working Group; Helander, *Havsörnen i Sverige*, 29.

32. Björn Helander, "Rapport till Statens Naturvårdsverk: Försök med artificiell inkubation av havsörnägg 1978," December 1978, Memorandums and Reports, Working Group.

33. Working Group, "Merikotkaprojektin projektiraportti," November 25, 1994, Letters 1976–1995, Working Group; Stjernberg, "Havsörnsskydd med avstamp," 43.

34. Stjernberg, "Havsörnsskydd med avstamp," 47.

35. Markku Lappalainen, "Talviruokinta on elvyttänyt merikotkakantaa," *Turun Sanomat*, February 16, 1992; "Merikotkia yhtä paljon kuin 1940-luvulla," *Helsingin Sanomat*, January 27, 1992.

36. See, for example, Meeting record, October 13, 1977; Meeting record, May 14, 1980; Meeting record, May 6, 1982, all in Records 1972–1986, Working Group; Press release, July 12, 1984, Nesting Reports; Press release, May 6, 1988, Letters 1976–1995; Juhani Koivusaari, "Merikotkien talviruokinta Merenkurkussa 1987–88," May 17, 1988, and "Merikotkien talviruokinta Merenkurkussa 1988–89," May 15, 1989, both in Nesting Reports, Working Group; "Talviruokinta kääntänyt merikotkakannan kasvuun," *Helsingin Sanomat*, April 21, 1991.

37. WWF Working Group, "Ilmoita merikotkahavainnosta—auta luonnonsuojelutyötä," Circular letter, n.d. [1980s], Various Documents, Working Group.

38. Quoted in Juhani Koivusaari, report on the upcoming nest inventories in the Kvarken area in 1984, August 2, 1983, n.p., Nesting Reports, Working Group. Also

see Seppo Ojala, report on eagles in Porttipahta, and flight inspections in Lapland, n.d. [1983], Nesting Reports; Ismo Nuuja to Jaakko Smolander, September 25, 1992, Research Unit, both in Working Group.

39. Henrik Wallgren, "Merikotkasuojelun organisatoriset puitteet," manuscript, 1993, Letters 1976–1995, Working Group.

40. *The White-Tailed Eagle Protection Strategy*, March 17, 1982, Memorandums and Reports; Juhani Koivusaari to the Environmental Protection Office of the Vaasa County Council, September 1, 1992, Research Unit; WWF press release, September 2, 1992, Research Unit; Juhani Koivusaari to Ulf Granqvist at the Vaasa Police Department, September 6, 1992, Research Unit, all in Working Group.

41. *The White-Tailed Eagle Protection Strategy*; Torsten Stjernberg, memorandum, August 28, 1992, Research Unit, Working Group.

42. See, for example, Henrik Wallgren and Pekka Borg to the Åland Provincial Government, December 3, 1980, Letters 1976–1995; Meeting record, April 13, 1992, Research Unit; Meeting record, March 1, 1993, Research Unit, all in Working Group; Nuuja, personal communication.

43. Henrik Wallgren, Torsten Stjernberg, and Lassi Karivalo, memorandum, March 7, 1988, Memorandums and Reports, Working Group.

44. For a more detailed analysis on the building of awareness on toxic substances in the Baltic marine environment, see Tuomas Räsänen, "Converging Environmental Knowledge: Re-evaluating the Birth of Modern Environmentalism in Finland," *Environment and History* 18 (2012): 159–81.

45. Henrik Wallgren, "Kaksikymmentä vuotta työtä merikotkien hyväksi—luonnonsuojelun menestystarina," draft letter to the Finnish Association for Nature Conservation, October 28, 1993, Letters, Working Group.

46. Puig de la Bellacasa, *Matters of Care*, 1.

47. See, for example, Krithika Srinivasan, "Caring for the Collective: Biopower and Agential Subjectification in Wildlife Conservation," *Environment and Planning D: Society and Space* 32 (2014): 506–8.

48. See, for example, Donna Haraway, *When Species Meet* (Minneapolis: University of Minnesota Press, 2008).

49. Haraway, *When Species Meet*.

CHAPTER 8: CAN YOU SHIP A PENGUIN COD?

1. J. W. Yates to Richard E. Byrd, November 7, 1939, Record Group 56.1, Series 1, subseries 2 (hereafter the series, colon, and subseries numbers, e.g., 1:2, all in the same record group), box 19, folder 738, Papers of Admiral Richard E. Byrd, Byrd Polar and Climate Research Center Archival Program, The Ohio State University, Columbus (hereafter Byrd Papers). Emphasis added.

2. "Penguin Requests," November 5 and 11, 1939, 1.2, box 19, folders 738, 739, 740, and 746, Byrd Papers. I am grateful to Laura Kissel for introducing me to this box, and her assistance throughout my visits to the archive.

3. "Byrd Wants a Home for His 20 Penguins," *Washington News*, November 7, 1939, newspaper clipping, 1.2, box 19, folder 742, Byrd Papers.

4. "Pet for Junior's Play Pen? Admiral Byrd Has Penguins," unattributed newspaper clipping, 1.2, box 19, folder 739, Byrd Papers.

5. Francis A. Gerstley, president of Whitman, to Byrd, telegram, November 6, 1939, 1.2, box 19, folder 738; Zeta of Alpha Kappa Lambda to Byrd, November 6, 1939, 1.2, box 19, folder 746, both in Byrd Papers.

6. Mustard Club, University of Michigan, to Byrd, November 6, 1939, 1.2, box 19, folder 746, Byrd Papers. On the Mustard Club, see "The Bowdoin Polar Bear," The Bowdoin Polar Bears, accessed November 14, 2022, http://bowdoin.prestosports.com/information/polarbear/index.

7. G. J. Guthrie Nicholson Jr. to Byrd, November 9, 1939, 1.2 box 19, folder 746, Byrd Papers.

8. On the intensity of Byrd's tour schedule, see Lisle Abbott Rose, *Explorer: The Life of Richard E. Byrd* (Columbia: University of Missouri Press, 2008), esp. chap. 6, "Celebrity."

9. Mr. George A. Brey, Jr, to Byrd, November 12, 1939; Burley Smith and William Bebler to Byrd, November 7, 1939; Arthur Maass to Byrd, November 6, 1939, all in folder 738; Goody Chambliss to Byrd, n.d., folder 739, all in 1.2, box 19, Byrd Papers.

10. Meredith Masters to Byrd, November 9, 1939, 1.2 box 19, folder 738, Byrd Papers.

11. Ann Lewis to Byrd, November 9, 1939, 1.2 box 19, folder 739, Byrd Papers.

12. Irvin D. Slusser to Byrd, telegram, November 6, 1939; Paul Lackie to Byrd, telegram, November 6, 1939; Unknown sender from Louisville, KY, telegram, n.d., all telegrams are filed in 1.2, box 19, folder 746, Byrd Papers.

13. Bruce School principal [no name] to Byrd, telegram, November 6, 1939, 1.2, box 19, folder 746, Byrd Papers.

14. Richard Carpenter to Byrd, November 6, 1939, 1.2, box 19, folder 740, Byrd Papers.

15. E. C. Appleby, "Mycosis of the Respiratory Tract in Penguins," *Proceedings of the Zoological Society of London* 139, no. 3 (October 1, 1962): 495–501, https://doi.org/10.1111/j.1469-7998.1962.tb01843.x.

16. Byrd to Henry and John North, October 24, 1939, 1.3, box 57, folder 2546, Byrd Papers.

17. "Sir Nils Olav," Edinburgh Zoo, accessed November 14, 2022, http://www.edinburghzoo.org.uk/sirnilsolav.

18. Peder Anker, *From Bauhaus to Ecohouse: A History of Ecological Design* (Baton Rouge: LSU Press, 2010), 19.

19. Roberts Peder and Dolly Jørgensen, "Animals as Instruments of Norwegian Imperial Authority in the Interwar Arctic," *Journal for the History of Environment and Society* 1, no. 1 (2016): 66.

20. Lewis L. Caton to Byrd, November 6, 1939, 1.2, box 19, folder 738, Byrd Papers.

21. Alice and Mildred Johnson to Byrd, November 7, 1939, 1.2, box 19, folder 738, Byrd Papers.

22. In 1903 they collected Adélie eggs, frozen, dead chicks, and one surviving chick. Ann Savours, *The Voyages of the Discovery: The Illustrated History of Scott's Ship* (London: Virgin, 1992), 4, 55.

23. Rose, *Explorer*, 286.

24. William J. Mills, s.v. "Siple, Paul," *Exploring Polar Frontiers: A Historical Encyclopedia* (Santa Barbara, CA: ABC-CLIO, 2003), 610.

25. Paul A. Siple and Alton A. Lindsey, "Ornithology of the Second Byrd Antarctic Expedition," *The Auk* 54, no. 2 (1937): 148.

26. "Byrd Party Off for Home after Year at Pole," *Chicago Daily Tribune*, February 7, 1935.

27. Marcia Winn, *Chicago Daily Tribune*, May 24, 1935.

28. "Another Polar Penguin Dies in Brookfield Zoo," *Chicago Tribune*, June 14, 1935.

29. "Habitat Group of Emperor Penguins at the Field Museum," *Science* 84, no. 2170 (July 31, 1936): 113.

30. Thelma Nurse to Byrd, November 6, 1939; Henry Burch to Byrd, November 7, 1939, both in 1.2, box 19, folder 740, Byrd Papers.

31. Gladys B. Erwin to Byrd, November 6, 1939, 1.2, box 19, folder 742, Byrd Papers.

32. Byrd to Leo MacDonald, May 15, 1939, 1.3, box 57, folder 2546, Byrd Papers.

33. Byrd to Trippe, n.d. [probably April 1939], 1.3, box 57, folder 2546, Byrd Papers.

34. "Byrd Lands 119 Penguins," *New York Times*, May 2, 1939.

35. "Byrd Lands 119 Penguins."

36. "Pet for Junior's Play Pen?"

37. Eleanor Wilson to Byrd, November 6, 1939, 1.2, box 19, folder 739, Byrd Papers.

38. Jack McGuire to Byrd, November 7, 1939, 1.2, box 19, folder 738, Byrd Papers.

39. Paul Turner to Byrd, November 5, 1939, 1.2, box 19, folder 738, Byrd Papers.

40. Stephen D. Perry, "CBS's Long Distance Radio Experiment: Broadcasting the Byrd Expedition from Little America," *Journal of Radio & Audio Media* 21, no. 1 (January 2014): 80, https://doi.org/10.1080/19376529.2014.891208.

41. On the technology involved, see "Listening to Byrd on Short-Wave Radio," RF Café, accessed November 14, 2022, http://www.rfcafe.com/references/radio-news/listen-byrd-july-1934-radio-news.htm.

42. Perry, "CBS's Long Distance Radio Experiment," 87, 90.

43. Danny Lewis, "A Brief History of Children Sent through the Mail," *Smithsonian Podcast*, accessed November 14, 2022, https://www.smithsonianmag.com/smart-news/brief-history-children-sent-through-mail-180959372/; "Object Spotlight: Collect on

Delivery," National Postal Museum, accessed November 14, 2022, https://postalmu seum.si.edu/collections/object-spotlight/collect-on-delivery-tag.

44. Nancy Pope, "Very Special Deliveries," National Postal Museum, accessed November 14, 2022, https://postalmuseum.si.edu/very-special-deliveries.

45. Nancy Pope, "100 Years of Parcels, Packages, and Packets, Oh My!" National Postal Museum, accessed November 14, 2022, https://postalmuseum.si.edu/research-articles/100-years-of-parcels-packages-and-packets-oh-my.

46. See John Hooton, "Frozen Fridays: 'L' Is for Little America!," The Ohio State University University Libraries," accessed November 14, 2022, https://library.osu.edu/blogs/archives/2017/02/03/frozen-fridays-l-is-for-little-america/.

47. "Object Spotlight: Antarctic Post Office," National Postal Museum, accessed November 14, 2022, https://postalmuseum.si.edu/collections/object-spotlight/antarctic-po.html.

48. Richard and Florence Atwater, *Mr. Popper's Penguins* (Boston: Little, Brown, 1992 [1938]), 14, 17.

49. Minutes of the board of directors meeting, June 15, 1928, Roll 2, folder 34, St. Louis Zoological Park Records 1910–1941, State Historical Society of Missouri, St. Louis. As we can't know how shipping was calculated (volume, weight, per animal?), this figure was established given that the cost to ship a penguin is relatively proportional to the cost to ship a sea elephant based on the purchase price for each animal. I thank my colleague Pamela Pyzza for help with this calculation.

50. "It's a Day Lost, but Byrd Still Has His Penguins," *Chicago Daily Tribune*, November 7, 1939.

51. Hazel McKercher to Joan Rigby, February 1940, 1.2, box 19, folder 748, Byrd Papers.

52. Minutes from the Zoological Board meetings, Roll 2, folders 34, 32, St. Louis Zoological Park Records 1910–1941.

53. Leo MacDonald to Henry Ringling Worth, October 16, 1939, 1.3, box 57, folder 2546, Byrd Papers.

54. Byrd to MacDonald, November 2, 1939, and November 3, 1939, both in 1.3, box 57, folder 2546, Byrd Papers.

55. Correspondence among McKercher, Byrd, MacDonald, and C. W. Coates, November 22, 1939–August 21, 1940, 1.3, box 61, folder 748, Byrd Papers.

56. Coates to McKercher, December 7, 1939, 1.3, box 61, folder 748, Byrd Papers.

57. McKercher to Coates, January 14, 1940, 1.3, box 61, folder 748, Byrd Papers.

58. Carl Heffenreffer to McKercher, January 14, 1940, 1.3, box 61, folder 748, Byrd Papers. Carl Haffenreffer, writing to update McKercher on the penguins' well-being, added that "since the purchase of these two birds our interest has, naturally, multiplied many fold, with the result that we have quite a Penguinana consisting of all types of pictures, statues, books, etc. etc."

59. Correspondence among Coates, MacDonald, and McKercher, November 22, 1939–August 21, 1940, 1.3, box 61, folder 748, Byrd Papers.

60. Coates to McKercher, February 5, 1940, 1.3, box 61, folder 748, Byrd Papers.

61. Newspaper clipping (no title), December 11, 1939, *New York Herald Tribune*, 1.2, box 19, folder 748, Byrd Papers.

62. Howard Rempes, "Pete Penguin," a summary of the penguin's life at the university, December 1939, material courtesy of Youngstown State University Archives & Special Collections, William F. Maag, Jr. Library, Youngstown State University, Youngstown, Ohio.

63. Howard Rempes, "Pete Penguin."

64. Marilyn Chuey, "An Historical Study of Varsity Football at Youngstown College" (master's thesis, Ohio University, 1956), 153. Sadly, Pete drowned in the pond he was kept while fishing under the ice on January 16, 1941.

65. Eileen Diamond to Byrd, 1.2, box 19, folder 738, Byrd Papers.

66. William Grub to Byrd, November 7, 1939; Doris Klutz to Byrd, November 7, 1939, both in 1.2, box 19, folder 738, Byrd Papers.

CHAPTER 9: SEEING THROUGH WINDOWS, WALLS, AND THE WEB IN PUBLIC AQUARIUMS

1. For a discussion of behind-the-scene technology in aquariums, see Leighton Taylor, *Aquariums: Windows to Nature* (New York: Prentice Hall General Reference, 1993), chap. 2.

2. Samantha Muka, *Oceans under Glass: Tank Craft and the Sciences of the Sea* (Chicago: University of Chicago Press, 2023).

3. Robert Jütte, *A History of the Senses: From Antiquity to Cyberspace*, trans. James Lynn (Cambridge: Polity, 2005).

4. Raymond Williams, *Television: Technology and Cultural Form* (New York: Schocken Books, 1975).

5. This was the third tenet in Dolly Jørgensen, "Not by Human Hands: Five Technological Tenets for Environmental History in the Anthropocene," *Environment & History* (2014): 479–89.

6. Bruno Latour, *Reassembling the Social: An Introduction to Actor-Network-Theory* (Oxford: Oxford University Press, 2005), 39.

7. Two collections of essays on undersea visual and literary aesthetics came out within a year of each other: Margaret Cohen and Killian Quigley, *The Aesthetics of the Undersea* (Abingdon, UK: Routledge, 2019); and Will Abberley, ed., *Underwater Worlds: Submerged Visions in Science and Culture* (Newcastle upon Tyne, UK: Cambridge Scholars, 2018). See also J. Malcolm Shick, "Toward an Aesthetic Marine Biology," *Art Journal* 67, no. 4 (2008): 62–86.

8. Ann Elias, *Coral Empire: Underwater Oceans, Colonial Tropics, Visual Modernity* (Durham, NC: Duke University Press, 2019); and Jonathan Christopher Crylen, "The Cinematic Aquarium: A History of Undersea Film" (PhD diss., University of Iowa, 2015).

9. Massimiliano Gaudiosi, "Marey's Aquarium: The Underwater World and the Archaeology of Cinema," in Abberley, *Underwater Worlds*, 185.

10. Judith Hamera, *Parlor Ponds: The Cultural Work of the American Home Aquarium* (Ann Arbor: University of Michigan Press, 2012), 24.

11. There had, however, long been fantasies of looking eye to eye with fish underwater. In a medieval romance written about Alexander the Great, there is a scene in which Alexander explores the ocean in a glass barrel. An excellent illustration of this story is found in a manuscript of the French prose Alexander romance from around 1420: Royal MS 20 B XX, f. 77v, British Library.

12. Jen Maylack, "How a Glass Terrarium Changed the World," *The Atlantic*, November 12, 2017, https://www.theatlantic.com/technology/archive/2017/11/how-a-glass-terrarium-changed-the-world/545621/.

13. Examples are Robert Warington, "Notice of Observations on the Adjustment of the Relations between the Animal and Vegetable Kingdoms, by Which the Vital Functions of Both Are Permanently Maintained," *Journal of Chemistry Society* 5 (1850): 52–54; and Philip Henry Gosse, "On Keeping Marine Animals and Plants Alive in Unchanged Sea Water," *Annual Magazine of Natural History*, 2nd ser., 10 (1852): 264. See Christopher Hamlin, "Robert Warington and the Moral Economy of the Aquarium," *Journal of the History of Biology* 19, no. 1 (1986): 131–53, for a discussion of the early use of aquariums in Victorian science.

14. Susan G. Davis, *Spectacular Nature: Corporate Culture and the Sea World Experience* (Berkeley: University of California Press, 1997): 99–100.

15. Elias, *Coral Empire*, esp. 156–59.

16. According to H. Noel Humphreys (*Ocean Gardens: The History of the Marine Aquarium* [London: Sampson Low, Son, 1857], 28), five thousand specimens passed through Gosse's hands en route to the Fish House.

17. Humphreys, *Ocean Gardens*, 27.

18. Humphreys, *Ocean Gardens*, 12.

19. The aquarium lighting from above at the Fish House was mentioned in Philip Lutley Sclater's *Guide to the Gardens of the Zoological Society of London*, 20th ed. (London: Bradbury, Evans, 1867), 43.

20. Philip Henry Gosse, *The Aquarium: An Unveiling of the Wonders of the Deep Sea* (London: John Van Voorst, 1854); and Shirley Hibberd, *The Book of the Aquarium and Water Cabinet; or Practical Instructions on the Formation, Stocking, and Management, in All Seasons of Collections of Fresh Water and Marine Life* (London: Groomsbridge & Sons, 1856). Shortly after came Henry D. Butler, *The Family Aquarium; or, Aqua Vivarium* (New York: Dick & Fitzgerald, 1858).

21. "L' aquarium: Vue intérieure," *Paris: Capital of the 19th Century*, 1860, Brown Digital Repository, Brown University Library, Providence, RI, https://repository.library.brown.edu/studio/item/bdr:86856/.

22. Andrew Flack, *The Wild Within: Histories of a Landmark British Zoo* (Charlottesville: University of Virginia Press, 2018), 8.

23. Flack, *Wild Within*, 8.

24. See, for example, Ian Jared Miller, *The Nature of Beasts: Empire and Exhibition at the Tokyo Imperial Zoo* (Oakland: University of California Press, 2021).

25. "Exhibition Aquariums," *The Aquarium* 3, no. 36 (July 1895): 182–84.

26. Elias, *Coral Empire*, 159.

27. See William R. Leach, *Land of Desire: Merchants, Power, and the Rise of a New American Culture* (New York: Vintage Books, 2011), 55–70, for this history.

28. Guillaume Le Gall, "Dioramas aquatiques: Théophile Gautier visite l'aquarium du Jardin d'acclimatation," *Culture & Musées* 32 (2018): 81–106. Sofie Lachapelle and Heena Mistry have also discussed the public aquarium at the Exposition Colonial Internationale (1931) in Paris as a panorama of empire: Lachapelle and Heena, "From the Waters of the Empire to the Tanks of Paris: The Creation and Early Years of the Aquarium Tropical, Palais de la Porte Dorée," *Journal of the History of Biology* 47 (2014): 1–27.

29. For discussions of the design of natural history dioramas, see Karen Wonders, "Habitat Dioramas and the Issue of Nativeness," *Landscape Research* 28, no. 1 (2003): 89–100; and Karen A. Rader and Victorian E. M. Cain, *Life on Display: Revolutionizing U.S. Museums of Science and Natural History in the Twentieth Century* (Chicago: University of Chicago Press, 2014).

30. Davis, *Spectacular Nature*, 100.

31. For descriptions of the facility, see "The New York Aquarium," *The Aquarium* 3, no. 35 (April 1895); John Treadwell Nichols, "The New York Aquarium," *The Aquarium* 1, no. 9 (February 1913): 75–76.

32. H. Dorner, *Guide to the New York Aquarium* (New York: Atheneum, 1877).

33. "Exhibition Aquariums."

34. Richard J. Conway, "The Detroit Aquarium," *The Aquarium* 2, no. 7 (December 1913): 66–67.

35. Butler, *Family Aquarium*, 23.

36. "Safetee Glass," *DuPont Magazine*, August 1923.

37. Ryoko has a short timeline of acrylic panel making on their website, accessed November 14, 2022, https://www.kkryoko.co.jp/en/products/acrytec/acrytec04.html.

38. Quoted in Mike Vogel, "New Aquariums Offer Fish-Eye View; Acrylic Panels Afford Exciting Look at Marine Ecosystems," *Buffalo News*, November 23, 1992.

39. Dennis Doordan, "Simulated Seas: Exhibition Design in Contemporary Aquariums," *Design Issues* 11, no. 2 (Summer 1995): 3–10.

40. For a discussion of the development of the Gulf of Mexico tanks in public aquariums and the ecosystems they show to visitors, see Dolly Jørgensen, "Mixing Oil and Water: Naturalizing Offshore Oil Platforms in American Aquariums," in *Oil Culture*, ed. D. Worden and R. Barrett, 267–88 (Minneapolis: University of Minnesota Press, 2014).

41. "Monterey Bay and Its Astonishing New Aquarium; It's a Good Time to Explore Both," *Sunset*, November 1, 1984.

42. Leighton Taylor, *Aquariums: Windows to Nature* (New York: Prentice Hall, 1993), 33.

43. Eva Hayward, "Sensational Jellyfish: Aquarium Affects and the Matter of Immersion," *differences: A Journal of Feminist Cultural Studies* 25, no. 3 (2012): 174.

44. The move of private aquariums from the home to the Web should not be unexpected. As Jody Berland has shown, the proliferation of animals in contemporary digital media is related to long-standing traditions of both exotic animal collection and commercialism with animals; Jody Berland, *Virtual Menageries: Animals as Mediators in Network Culture* (Cambridge, MA: MIT Press, 2019).

45. S. Mary P. Benbow, "A View through the Glass: Aquariums on the Internet," *Internet Research* 7, no. 1 (1997): 27–31.

46. Early views of the Fish Cam can be seen on the Wayback Machine, accessed April 28, 2024, https://web.archive.org/web/20240000000000/http://wp.netscape .com/fishcam/fishcam_about.html.

47. "Pausefiskene," NRK, October 16, 2009, https://www.nrk.no/kultur/pause fiskene-1.6786800. A pausefisk broadcast from January 1, 1972, is available to watch on the NRK website: accessed November 14, 2022, https://tv.nrk.no/program/ FRED72000472.

48. "Live Cams," Monterey Bay Aquarium, accessed November 14, 2022, https:// www.montereybayaquarium.org/animals/live-cams. In 2014 the Kelp Forest Cam and Open Sea Cam were the only underwater views available on the website, according to an archived version of the page; the others have been added since then.

49. The webcam feeds are available at Aquarium of the Pacific, accessed November 14, 2022, https://explore.org/livecams/aquarium-of-the-pacific/. Webcams were first made available on the aquarium's website in July 2012, accessed November 14, 2022, https://web.archive.org/web/20120718192255/http://www.aquariumofpacific .org:80/exhibits/webcams/.

50. "Birch Aquarium at Scripps Unveils Live HD Web Cam of Its Two-Story Kelp Forest," Scripps Institution of Oceanography, via the Wayback Machine, accessed November 14, 2022, https://web.archive.org/web/20150910154101/https://scripps.ucsd .edu/news/1992.

51. This is different from the earliest underwater photography, which required physical separation of the camera equipment from the water, through technologies such as the photosphere, as discussed in Elias, *Coral Empire*.

52. John Berger, *Why Look at Animals?* (London: Penguin Books, 2009), 27.

53. Sumida Aquarium, "Please Remember about Humans," accessed April 28, 2024, https://www.sumida-aquarium.com/news/details/2236/; Sumida Aquarium (@Sumida_Aquarium), tweet, accessed April 28, 2024, https://x.com/Sumida _Aquarium/status/1257552318033063938.

54. Finn Arne Jørgensen, "The Internet Is Obsessed with a Video Feed of Bears Eating Salmon," *The Atlantic*, August 22, 2016, https://www.theatlantic.com/technology/ archive/2016/08/bear-cam/495638/.

55. See Gregg Mitman, *Reel Nature: America's Romance with Wildlife on Film* (Madison: University of Wisconsin Press, 2009), for a discussion of how new technologies and scientific developments shaped encounters with wildlife through film.

56. Jonathon Turnbull, Adam Searle, and William M. Adams, "Quarantine Encounters with Digital Animals: More-Than-Human Geographies of Lockdown Life," *Journal of Environmental Media* 1, suppl. (2020): 6.7.

57. For a fascinating analysis of how humans have tried to adapt themselves through technology to aquatic life, see Helen M. Rozwadowski, "'Bringing Humanity Full Circle Back into the Sea': *Homo aquaticus*, Evolution, and the Ocean," *Environmental Humanities* 14, no. 1 (2022): 1–28.

CHAPTER 10: TOMORROW'S MORE-THAN-HUMAN MINERS OF THE DEEP

1. Pradeep A. Singh, "Deep Seabed Mining and Sustainable Development Goal 14," in *Life below Water*, ed. Walter Leal Filho, Anabela Marisa Azul, Luciana Brandli, Amanda Lange Salvia, and Tony Wall (Cham, Switzerland: Springer International, 2021), 1–13.

2. Lauren Rickards, "Metaphor and the Anthropocene: Presenting Humans as a Geological Force," *Geographical Research* 53, no. 3 (2015): 280–87.

3. Donna Haraway, "Anthropocene, Capitalocene, Plantationocene, Chthulucene: Making Kin," *Environmental Humanities* 6 (2015): 159–65.

4. Tomas Frederiksen and Matthew Himley, "Tactics of Dispossession: Access, Power, and Subjectivity at the Extractive Frontier," *Transactions of the Institute of British Geographers* 45, no. 1 (2020): 50–64.

5. Kate D. Derickson, "Urban Geography I: Locating Urban Theory in the 'Urban Age,'" *Progress in Human Geography* 39, no. 5 (2015): 653.

6. Hillary Angelo and David Wachsmuth, "Urbanizing Urban Political Ecology: A Critique of Methodological Cityism," *International Journal of Urban and Regional Research* 39, no. 1 (2015): 16–27; Neil Brenner, "Urban Theory without an Outside," *Harvard Design Magazine*, no. 37 (2014): 42–47; Andy Merrifield, "The Urban Question under Planetary Urbanisation," in *Implosions/Explosions: Towards a Study of Planetary Urbanization*, ed. Neil Brenner. 164–80 (Berlin: Jovis, 2014); Lauren Rickards, Brendan Gleeson, Mark Boyle, and Cian O'Callaghan, "Urban Studies after the Age of the City," *Urban Studies* 53, no. 8 (2016): 1523–41.

7. Neil Brenner and Nik Theodore, "Cities and the Geographies of 'Actually Existing Neoliberalism,'" in *Critique of Urbanization: Selected Essays*, ed. Brenner, 42–68 (Basel: Birkhäuser, 2017 [2002]); Brenner and Christian Schmid, "Planetary Urbanisation," in *Urban Constellations*, ed. Matthew Gandy, 10–13 (Berlin: Jovis Verlag, 2011).

8. Brenner and Theodore, "Cities and the Geographies of 'Actually Existing Neoliberalism'"; William Cronon, "The Trouble with Wilderness; or, Getting Back to the Wrong Nature," in *Uncommon Ground: Rethinking the Human Place in Nature*, ed. Cronon, 69–90 (New York: W. W. Norton, 1995).

9. Brenner, "Urban Theory without an Outside."

10. For the extension of urbanization beyond cities, see Brenner, "Urban Theory without an Outside"; Brenner and Schmid, "Planetary Urbanisation"; and Rickards et al., "Urban Studies after the Age of the City." On oceanic manifestations of this argument, see Ross Exo Adams, "Mare Magnum: Urbanization of the Sea," in *Association of American Geographers Conference 2015* (Chicago: AAG, 2015); Martín Arboleda and Daniel Banoub, "Market Monstrosity in Industrial Fishing: Capital as Subject and the Urbanization of Nature," *Social & Cultural Geography* (2016): 1–19; and Nancy Couling and Carola Hein, eds., *The Urbanisation of the Sea: From Concepts and Analysis to Design* (Rotterdam: nai 010, 2020).

11. Philip E. Steinberg, "Sovereignty, Territory, and the Mapping of Mobility: A View from the Outside," *Annals of the Association of American Geographers* 99, no. 3 (2009): 467–95; Henry Jones, "Lines in the Ocean: Thinking with the Sea about Territory and International Law," *London Review of International Law* 4, no. 2 (2016): 307–43.

12. Henri Lefebvre, *The Urban Revolution* (Minneapolis: University of Minnesota Press, 2003 [1970]), 29.

13. Neil Smith, foreword to *The Urban Revolution*, by Henri Lefebvre (Minneapolis: University of Minnesota Press, 2003).

14. Henri Lefebvre, "Right to the City," in *Writings on Cities*, ed. Eleonore Kofman and Elizabeth Lebas (Oxford: Blackwell, 1996 [1968]), 143.

15. Alan D. Hemmings, Klaus Dodds, and Peder Roberts, "Introduction: The Politics of Antarctica," in *Handbook on the Politics of Antarctica*, ed. Dodds, Hemmings, and Roberts, 1–20 (Cheltenham, UK: Edward Elgar, 2017).

16. Alessandro Antonello, "Life, Ice and Ocean: Contemporary Antarctic Spaces," in Dodds, Hemmings, and Roberts, *Handbook on the Politics of Antarctica*.

17. Johan Rockström et al., "Planetary Boundaries: Exploring the Safe Operating Space for Humanity," *Ecology and Society* 14, no. 2 (2009): n.p. "Spaceship Earth" was coined by R. Buckminster Fuller in *Operating Manual for Spaceship Earth* (New York: Simon and Schuster, 1969).

18. Olivier Vidal, Bruno Goffé, and Nicholas Arndt. "Metals for a Low-Carbon Society," *Nature Geoscience* 6, no. 11 (2013): 894–96.

19. T. Kuhn, A. Wegorzewski, C. Rühlemann, and A. Vink, "Composition, Formation, and Occurrence of Polymetallic Nodules," in *Deep-Sea Mining: Resource Potential, Technical and Environmental Considerations*, ed. Sharma Rahul, 23–63 (Cham, Switzerland: Springer International, 2017).

20. Jack N. Barkenbus, *Deep Seabed Resources: Politics and Technology* (New York: Free Press, 1979); International Seabed Authority, *Technical Study: Polymetallic Nodules* (Kingston, Jamaica: International Seabed Authority, 2008), www.isa.org.jm/files/doc uments/EN/Brochures/ENG7.pdf; Nico Schrijver, *Sovereignty over Natural Resources: Balancing Rights and Duties* (Cambridge: Cambridge University Press, 1997).

21. M. Atmanand and G. Ramadass, "Concepts of Deep-Sea Mining Technologies," in Rahul, *Deep-Sea Mining*.

22. Robert Bogue, "Underwater Robots: A Review of Technologies and Applications," *Industrial Robot* 42, no. 3 (2015): 186–91.

23. Robert D. Christ and Robert L. Wernli Sr., *The ROV Manual: A User Guide for Remotely Operated Vehicles* (Oxford: Elsevier Science, 2013).

24. Fu-dong Gao, Yan-yan Han, Hai-dong Wang, and Nan Xu, "Analysis and Innovation of Key Technologies for Autonomous Underwater Vehicles," *Journal of Central South University* 22 (2015): 3347–57.

25. Paul S. Grad, "Running with Robotics," *Engineering & Mining Journal* 211, no. 1 (2010): 34–36.

26. Kevin Rawlinson, "Search for Shackleton's Endurance Called Off after Loss of Submarine," *The Guardian*, February 15, 2019, Australia ed., https://www.the guardian.com/world/2019/feb/14/search-for-shackletons-endurance-called-off-after -loss-of-submarine.

27. John Shears, Julian Dowdeswell, Freddie Ligthelm, and Paul Wachter, "The Weddell Sea Expedition 2019," *EGU General Assembly 2020*, May 4–8, 2020, https:// doi.org/10.5194/egusphere-egu2020-21896.

28. Matt Simon, "Dive under the Ice with the Brave Robots of Antarctica," *Wired*, July 19, 2018, https://www.wired.com/story/dive-under-the-ice-with-the-brave-robots -of-antarctica/.

29. Fabien Roquet, Guy Williams, Mark A. Hindell, Rob Harcourt, Clive McMahon, Christophe Guinet, Jean-Benoit Charrassin, et al., "A Southern Indian Ocean Database of Hydrographic Profiles Obtained with Instrumented Elephant Seals," *Scientific Data* 1 (2014): 140028.

30. Mark A. Hindell, Clive R. McMahon, Marthán N. Bester, Lars Boehme, Daniel Costa, Mike A. Fedak, Christophe Guinet, et al., "Circumpolar Habitat Use in the Southern Elephant Seal: Implications for Foraging Success and Population Trajectories," *Ecosphere* 7, no. 5 (2016): 4.

31. L. Boehme, P. Lovell, M. Biuw, F. Roquet, J. Nicholson, S. E. Thorpe, M. P. Meredith, and M. Fedak, "Technical Note: Animal-Borne CTD-Satellite Relay Data Loggers for Real-Time Oceanographic Data Collection," *Ocean Science* 5, no. 4 (2009): 692.

32. Jody Emel and Jennifer Wolch, "Witnessing the Animal Moment," in *Animal Geographies: Place, Politics, and Identity in the Nature-Culture Borderlands*, ed. Emel and Wolch (London: Verso, 1998).

33. An uncanny category under which we may well register augmented marine mammals such as the southern elephant seal.

34. Chris Russell, "Is the Earth a Medium? Situating the Planetary in Media Theory," *Ctrl-Z* 7 (2017), http://www.ctrl-z.net.au/articles/issue-7/russill-is-the-earth -a-medium/.

35. Chris Philo, "(In)secure Environments and the Domination of Nature," *Geographical Journal* 181 (2015): 322–37.

36. Ruth Panelli, "More-Than-Human Social Geographies: Posthuman and Other Possibilities," *Progress in Human Geography* 34, no. 1 (2010): 85.

37. Jane Bennett, *Vibrant Matter: A Political Ecology of Things* (Durham, NC: Duke University Press, 2010).

38. Anna Lowenhaupt Tsing, "Strathern beyond the Human: Testimony of a Spore," *Theory, Culture & Society* 31, no. 2–3 (2014): 223.

39. Tsing, "Strathern beyond the Human," 224.

40. Tsing, "Strathern beyond the Human," 224.

41. Derickson, "Urban Geography I."

42. Jessica S. Lehman, "Volumes beyond Volumetrics: A Response to Simon Dalby's 'The Geopolitics of Climate Change,'" *Political Geography* 37 (2013): 51–52.

43. Donna J. Haraway, *Staying with the Trouble: Making Kin in the Chthulucene* (Durham, NC: Duke University Press, 2016), 48.

44. Jonathan Crary, *Techniques of the Observer: On Vision and Modernity in the Nineteenth Century* (Cambridge, MA: MIT Press, 1990).

45. Paul Virilio, *The Vision Machine* (London: British Film Institute, 1994 [1988]).

46. Tsing, "Strathern beyond the Human," 224.

47. Bennett, *Vibrant Matter*, 56.

48. Bennett, *Vibrant Matter*, 58.

49. A. Knobloch and T. Kuhn, "Predictive Mapping of the Nodule Abundance and Mineral Resource Estimation in the Clarion-Clipperton Zone Using Artificial Neural Networks and Classical Geostatistical Methods," in Rahul, *Deep-Sea Mining*.

50. Gavin M. Mudd, "Gold Mining in Australia: Linking Historical Trends and Environmental and Resource Sustainability." *Environmental Science & Policy* 10, no. 7 (2007): 636.

51. Donna J. Haraway, *Simians, Cyborgs, and Women: The Reinvention of Nature* (New York: Routledge, 1988); Bill Brown, "Thing Theory," *Critical Inquiry* 28, no. 1 (2001), 1–22.

52. Natalie Klein, Douglas Guilfoyle, Md. Saiful Karim, and Rob McLaughlin, "Maritime Autonomous Vehicles: New Frontiers in the Law of the Sea," *International and Comparative Law Quarterly* 69, no. 3 (2020): 719–34; Rob Harcourt, Ana M. M. Sequeira, Xuelei Zhang, Fabien Roquet, Kosei Komatsu, Michelle Heupel, Clive McMahon, et al., "Animal-Borne Telemetry: An Integral Component of the Ocean Observing Toolkit," *Frontiers in Marine Science* 6, no. 326 (2019): 1–21.

CHAPTER 11: GUIDE DOGS AND ELECTRONIC MOBILITY AIDS AS THE MEDIATORS OF ACCESSIBILITY

This research has been supported by the projects PRG314, Semiotic Fitting as a Mechanism of Biocultural Diversity: Instability and Sustainability in Novel Environments, and PRG 1504, Meanings of Endangered Species in Culture: Ecology, Semiotic Modelling and Reception.

1. Assistive Technology Industry Association, *What Is AT?*, accessed August 24, 2021, https://www.atia.org/home/at-resources/what-is-at/.

2. The European Pet Food Industry, *Annual Report, 2021*, 2021, https://www.atia.org/home/at-resources/what-is-at/.

3. The term "Umwelt" was introduced by the Baltic German biologist Jakob von Uexküll (1864–1944) to denote the species-specific subjective world of the animal, encompassing those parts of the environment that the animal is capable of perceiving and acting upon; von Uexküll, *Umwelt und Innenwelt der Tiere* (Berlin: Springer, 1909).

4. Nicholas A. Giudice and Gordon E. Legge, "Blind Navigation and the Role of Technology," in *The Engineering Handbook of Smart Technology for Aging, Disability, and Independence*, ed. Abdelsalam Helal, Mounir Mokhtari, and Bessam Abdulrazak, 479–500 (Hoboken, NJ: Wiley, 2008).

5. Shlomo Deshen and Hilda Deshen, "On Social Aspects of the Usage of Guide Dogs and Long-Canes," *Sociological Review* 37, no. 1 (1989): 89–103.

6. Wafa Elmannai and Khaled Elleithy, "Sensor-Based Assistive Devices for Visually-Impaired People: Current Status, Challenges, and Future Directions," *Sensors* 17, no. 3 (2017): 1–42; Alexy Bhowmick and Shyamanta M. Hazarika, "An Insight into Assistive Technology for the Visually Impaired and Blind People: State-of-the-Art and Future Trends," *Journal of Multimodal User Interfaces* 11, no. 2 (2017): 149–72; Dimitrios Dakopoulos and Nikolaos G. Bourbakis, "Wearable Obstacle Avoidance Electronic Travel Aids for Blind: A Survey," *IEEE Transactions on Systems, Man and, Cybernetics—Part C: Applications and Reviews* 40, no. 31 (2010): 25–34.

7. Marion A. Hersh and Michael A. Johnson, "Mobility: An Overview," in *Assistive Technology for Visually Impaired and Blind People*, ed. Hersh and Johnson, 167–208 (London: Springer, 2008).

8. Serena Zanolla, "Interactive and Multimodal Environments for Learning" (PhD diss., University of Udine, 2013), 59; Pawel Strumillo, "Electronic Interfaces Aiding the Visually Impaired in Environmental Access, Mobility and Navigation," in *3d Conference on Human System Interactions (HSI)*, ed. Tomasz Pardela, Bogdan Wilamowski, 17–24 (Rzeszow, Poland: IEEE-Press, 2010).

9. Giudice and Legge, "Blind Navigation and the Role of Technology," 485.

10. Strumillo, "Electronic Interfaces Aiding the Visually Impaired," 20.

11. Zanolla, "Interactive and Multimodal Environments for Learning," 60.

12. Umme Kawsar Alam, Fazle Rabby, and M. T. Islam, "Development of a Technical Device Named GPS Based Walking Stick for the Blind," *Rajshahi University Journal of Science & Engineering* 43 (2015): 73–80.

13. Giudice and Legge, "Blind Navigation and the Role of Technology," 485.

14. Julie Anderson and Neil Pemberton, "Walking Alone: Aiding the War and

Civilian Blind in the Inter-war Period," *European Review of History—Revue Européenne d'Histoire* 14, no. 4 (2007): 459–79.

15. Anderson and Pemberton, "Walking Alone," 463.

16. Monika Baar, "Prosthesis for the Body and for the Soul: The Origins of Guide Dog Provision for Blind Veterans in Interwar Germany," *First World War Studies* 6, no. 1 (2015): 81–98.

17. Emanuel Sarris, "Ein neues Verfahren: Führhunde für Blinde auszubilden," *Der Naturforscher* 12 (1935): 260–65.

18. The interviews were conducted from winter 2010–2011 to spring 2014 and included thirty-six phone and face-to-face interviews with guide dog users from different European countries (focus groups being in Sweden, Germany, and Estonia); see Riin Magnus, "The Semiotic Challenges of Guide Dog Teams: The Experiences of German, Estonian and Swedish Guide Dog Users," *Biosemiotics* 9, no. 2 (2015): 267–85.

19. For the choice pertaining to guide dogs, see Lorraine E. Whitmarsh, "The Benefits of Guide Dog Ownership," *Visual Impairment Research* 7, no. 1 (2005): 27–42.

20. The term "affordance," initially introduced by James Gibson, refers to an element in the environment that allows certain activities for organisms with certain bodily structures; James Gibson, *Ecological Approach to Visual Perception* (New York: Taylor & Francis Group, 1986). It can be seen as an equivalent of Umwelt in establishing functional relations with the environment. Although both are relational terms (the entanglement of organism and environment being presumed), the term "Umwelt" stresses more the organism's semiotic abilities, whereas the term "affordance" points to the environmental ground for certain activities.

21. Donna Haraway, *Staying with the Trouble: Making Kin in the Chthulucene* (Durham, NC: Duke University Press, 2016).

22. Michele Merritt, *Minding Dogs: Humans, Canine Companions, and a New Philosophy of Cognitive Science* (Athens: University of Georgia Press, 2021).

23. Merritt, *Minding Dogs*, 3.

24. Brian Hare and Michael Tomasello, "Human-like Social Skills in Dogs?" *Trends in Cognitive Sciences* 9, no. 9 (2005): 439–44.

25. Charlotte Duranton and Florence Gaunet, "*Canis sensitivus*: Affiliation and Dogs' Sensitivity to Others' Behavior as the Basis for Synchronization with Humans?," *Journal of Veterinary Behavior: Clinical Applications and Research* 10, no. 6 (2015): 513–24; Charlotte Duranton, Thierry Bedossa, and Florence Gaunet, "Interspecific Behavioural Synchronization: Dogs Exhibit Locomotor Synchrony with Humans," *Scientific Reports* 7, no. 1 (2017); Duranton, Bedossa, and Gaunet, "When Facing an Unfamiliar Person, Pet Dogs Present Social Referencing Based on Their Owner's Direction of Movement Alone," *Animal Behaviour* 113 (2016): 147–56.

26. Charlotte Duranton and Florence Gaunet, "Behavioural Synchronization from an Ethological Perspective: Overview of Its Adaptive Value," *Adaptive Behavior* 24, no. 3 (2016): 181–91.

27. Duranton, Bedossa, and Gaunet, "Interspecific Behavioural Synchronization."

28. Szima Naderi, Adam Miklósi, Antal Dóka, and Vilmos Csányi, "Co-operative Interactions between Blind Persons and Their Dogs," *Applied Animal Behaviour Science* 74, no. 1 (2001): 59–80.

29. Gregory Bateson, *Steps to an Ecology of Mind* (Chicago: University of Chicago Press, 2000).

30. Giudice and Legge, "Blind Navigation and the Role of Technology"; Dakopoulos and Bourbakis, "Wearable Obstacle Avoidance Electronic Travel Aids for Blind."

31. See Giudice and Legge, "Blind Navigation and the Role of Technology."

32. Iconic signs establish a relation between the signifier and signified through similarity, indexical signs through contingency, and symbolic signs through an arbitrary link.

33. James Gibson, *Ecological Approach to Visual Perception* (New York: Taylor & Francis Group, 1986).

34. See, e.g., William Warren, "Constructing an Econiche," in *Global Perspectives on the Ecology of Human-Machine Systems*, ed. John M. Flach, Peter A. Hancock, Jeff Caird and Kim J. Vicente, 210–37 (Hilldsale, NJ: Erlbaum, 1995); Brian Zaff, "Designing with Affordances in Mind," both in Flach et al., *Global Perspectives on the Ecology of Human-Machine Systems*; Jonathan R. A. Maier and Georges M. Fadel, "An Affordance-Based Approach to Architectural Theory, Design, and Practice," *Design Studies* 30, no. 4 (2009): 393–414.

35. Dakopoulos and Bourbakis, "Wearable Obstacle Avoidance Electronic Travel Aids for Blind."

36. Gibson, *Ecological Approach to Visual Perception*, 34.

37. Warren, "Constructing an Econiche."

38. Don Norman, *The Design of Everyday Things* (New York: Basic Books, 2013), 14.

39. See Emanuel Sarris, "Die Befähigung des Hundes," *Umschau* 38 (1934): 106–10.

40. Gibson, *Ecological Approach to Visual Perception*, 34.

41. Gibson, *Ecological Approach to Visual Perception*, 135.

42. Vasilis Galis, "We Have Never Been Able-Bodied," in *Routledge Handbook of Disability Studies*, ed. Nick Watson, Alan Roulstone, and Carol Thomas (London: Routledge, 2019), 411.

43. For the complex of challenges guide dog users face, see Riin Magnus, "The Semiotic Challenges of Guide Dog Teams: The Experiences of German, Estonian and Swedish Guide Dog Users," *Biosemiotics* 9 (2016): 267–85.

44. Donna Haraway, *When Species Meet* (Minneapolis: University of Minnesota Press, 2008).

45. Vasilis Galis, "Enacting Disability: How Can Science and Technology Studies Inform Disability Studies?," *Disability & Society* 26, no, 7 (2011): 825–38; Ingunn Moser, "Disability and the Promises of Technology: Technology, Subjectivity and Embodiment within an Order of the Normal," *Information, Communication & Society* 9 (2006): 373–95.

46. Sunaura Taylor, *Beasts of Burden: Animal and Disability Liberation* (New York: New Press, 2017), 14.

47. Moser, "Disability and the Promises of Technology," 380.

48. Thom van Dooren, Eben Kirskey, and Ursula Münster, "Multispecies Studies Cultivating Arts of Attentiveness," *Environmental Humanities* 8, no. 1 (2016): 1–22.

CHAPTER 12: THE HUMAN-CAMERA-ANIMAL TECHNOLOGICAL LOVE TRIANGLE

This chapter was finished during a Juan de la Cierva-Incorporación postdoctoral fellowship funded by the Spanish Ministry of Science and Innovation. I would also like to thank the four anonymous reviewers whose comments benefited the final result.

1. Consider "Closing the Gap," the title of the workshop that originated this book. "Gap" has several meanings: among others, separation or absence, as well as temporal and spatial connotations. The image that the phrase brings to my mind is that of a bridge being built, or the process of filling up a breach or void with solid matter. In short, a matter of distance, the act of connecting two previously isolated points or locations, of bringing two shores or borders together.

2. I addressed this as part of the online workshop Animal / Privacy: Historical and Conceptual Approaches, November 8–9, 2021, https://animalprivacy.wordpress .com/.

3. I am using "orthopedic" here to portray cameras and the accompanying technology as an aid that is mending, although not necessarily in an adequate manner, the strained relationships between humans and other animals.

4. Elsewhere I have focused mainly on the physical and metaphorical dimensions, and on animals who steal cameras or photobomb humans: Concepción Cortés Zulueta, "That Seagull Stole My Camera (and My Shot)! Overlapping Metaphorical and Physical Distances in the Human-Animal-Camera Triad," in *Critical Distance in Documentary Media*, ed. Gerda Cammaer, Blake Fitzpatrick, and Bruno Lessard, 231–56 (Cham, Switzerland: Palgrave Macmillan, 2018).

5. "A Modern Jonah," *San Francisco Call*, May 11, 1891, 8, https://cdnc.ucr.edu/cgi -bin/cdnc?a=d&d=SFC18910511.2.131. I haven't found the *New York Ledger* report of the incident, but almost the same article appears, under a different title, "Stories about Whales," in the *Los Angeles Herald*, July 21, 1907.

6. Walter B. Cannon, *Bodily Changes in Pain Hunger Fear and Rage* (New York: D. Appleton, 1927); Cannon, "The James-Lange Theory of Emotions: A Critical Examination and an Alternative Theory," *American Journal of Psychology* 100, no. 3–4 (Fall-Winter 1987): 567–86.

7. For instance, see the background section in Norman B. Schmidt et al., "Exploring Human Freeze Responses to a Threat Stressor," *Journal of Behavior Therapy and Experimental Psychiatry* 39, no. 3 (September 2008): 292–304.

8. One example is a vicious dog in Leon F. Seltzer, "Trauma and the Freeze Response:

Good, Bad, or Both?," *Psychology Today*, July 2015, https://www.psychologytoday.com/blog/evolution-the-self/201507/trauma-and-the-freeze-response-good-bad-or-both.

9. I find this initial sentence a bit optimistic for the late nineteenth century (most people really knew?), considering the bad reputation that was associated with cetaceans through whaling shipwrecks and Moby-Dick-like literature extending well into the twentieth century.

10. A possibility is that the bumps and persistence on the side of the whale could have come from past bad experiences with humans, whaling having been widespread and particularly cruel at the time. Since then human and whale cultures (and memories) have changed. Sometime after this chapter was written a family of another kind of whale, orcas, began to interact with the rudders of some of the vessels they encountered in the Strait of Gibraltar region, altering the boats' courses or even leaving them adrift. The reasons and motivations behind this cultural behavior of this family, which has been named Gladis, are still being studied, but orcas have often been described as being driven by curiosity. Both the behavior itself and the reactions to it in social media, including hashtags like #teamorca and viral memes that hail these whales as antifascist, anti-rich, or communist heroes, would make for another chapter.

11. This is in spite of certain attempts at portraying as attacks other exploratory or regular behaviors: "Terrifying Video Shows Moment a Pair of Divers Were Almost Eaten by Two Hungry Humpback Whales," *Daily Mail Online*, July 23, 2013, https://www.dailymail.co.uk/news/article-2374882/Hungry-humpback-whales-video-shows-pair-divers-eaten.html.

12. Brandon Spektor, "This Humpback Whale Saved a Woman's Life, but Probably Not on Purpose," Live Science, January 9, 2018, https://www.livescience.com/61380-humpback-whale-saves-diver-video.html. Also significant is the scientific insistence on underscoring possible unintentionality and self-interest instead of potential altruism.

13. Reinhard Döllner, *Watching, Touching and Kissing Whales in Laguna San Ignacio / Mexico*, 2016, https://www.youtube.com/watch?v=I8ecwgaPMks; also available in an original version with German commentary, https://www.youtube.com/watch?v=YlR-D713qY0.

14. There would be much to write about on the mix of languages in the video and the social and economic dynamics involved in activities like whale watching, as there is a clear North–South axis regarding where the whale watchers come from, and where they usually travel to engage in the activity.

15. Another relevant issue that I cannot thoroughly address here is how attitudes to whales shifted in the 1960s and 1970s. See D. Graham Burnett, *The Sounding of the Whale: Science & Cetaceans in the Twentieth Century* (Chicago: University of Chicago Press, 2013). I discussed this topic in Concepción Cortés Zulueta, *Fundamentos biológicos de la creación: Animales en el arte y arte animal*, vol. 1 (Madrid: Universidad Autónoma de Madrid, 2016), 1:513–42, http://hdl.handle.net/10486/672482.

16. Susan Sontag, *On Photography* (New York: Farrar, Straus and Giroux, 1977), 13.

17. Joan Fontcuberta, *La furia de las imágenes: Notas sobre la postfotografía* (Barcelona: Galaxia Gutenberg, 2016).

18. "Muere una cría de delfín en Almería por el acoso de los bañistas," *Público*, August 15, 2017, https://www.publico.es/sociedad/muere-cria-delfin-almeria-acoso-banistas .html.

19. Toyin Owoseje, "Tourist Trampled to Death by Elephant in Zimbabwe as She Tried to Take Photo," *The Independent*, September 27, 2018, https://www.inde pendent.co.uk/news/world/elephant-tramples-german-tourist-death-zimbabwe-mana -pool-a8557561.html; "Hippo Bite Kills Tourist in Kenya," *BBC News*, August 13, 2018, sec. Africa, https://www.bbc.com/news/world-africa-45162747. Again, the North–South dynamic is very present in these cases.

20. I think he was a male, but I might be wrong.

21. Chad Pawson, "Sea Lion Drags Girl Underwater in Richmond, B.C.," *CBC News*, May 21, 2017, https://www.cbc.ca/news/canada/british-columbia/sea-lion-drags -girl-into-water-off-steveston-docks-in-richmond-b-c-1.4125848.

22. Brian Huang et al., "Take Only Pictures, Leave Only . . . Fear? The Effects of Photography on the West Indian Anole Anolis Cristatellus," *Current Zoology* 57, no. 1 (2011): 77–82.

23. International Whaling Commission, "General Principles for Whalewatching," accessed November 1, 2018, https://iwc.int/wwguidelines.

24. For instance, the Canadian regulations are illustrated in a very visual and clear way: Fisheries and Oceans Canada, "Watching Marine Wildlife," accessed November 1, 2018, http://www.dfo-mpo.gc.ca/species-especes/mammals-mammiferes/watching -observation/index-eng.html.

25. This means that the contacts and interactions seen in the Laguna San Ignacio video are not allowed or sanctioned in Canada, while in Mexico the regulations establish that you have to wait for the departure of the whale that has approached your boat, and the only interactions that are prohibited are ones that interfere with the natural behavior of the whales or involve a forced physical contact: "NORMA Oficial Mexicana NOM-131-SEMARNAT-2010, que establece lineamientos y especificaciones para el desarrollo de actividades de observación de ballenas, relativas a su protección y la conservación de su hábitat," *Diario Oficial de la Federación—DOF*, October 17, 2011, http://www.dof.gob.mx/nota_detalle.php?codigo=5214459&fecha=17/10/2011; Fisheries and Oceans Canada, "Watching Marine Wildlife"; Public Works and Government Services, Government of Canada, "Regulations Amending the Marine Mammal Regulations: SOR/2018–126," *Canada Gazette*, July 11, 2018, http://www.gazette .gc.ca/rp-pr/p2/2018/2018-07-11/html/sor-dors126-eng.html.

26. Sontag, *On Photography*, 8–9.

27. This chapter was initially written well before the COVID-19 pandemic, but it feels like the lockdowns experienced around the globe accelerated and intensified the bonds with these nonhuman animal members of our virtual communities, featured in millions of images shared worldwide, while some of their tangible counterparts roamed

the streets, empty of humans. I believe this was elegantly pointed out in the non-exploding and special opening credits of the first chapter of *The Good Fight*'s season five, full of cute and soft creatures, once this remarkable TV series was resumed after the first wave of the pandemic.

28. Other ruptures of cushioned scripts in another kind of human-camera-animal encounter take place on nest cams, for example, when chicks are not well fed. Indeed, it is not uncommon that viewers demand interventions to mend the plot: Karin Brulliard, "People Love Watching Nature on Nest Cams—Until It Gets Grisly," *Washington Post*, May 19, 2016, https://www.washingtonpost.com/news/animalia/wp/2016/05/19/when-nest-cams-get-gruesome-some-viewers-cant-take-it/.

29. Fontcuberta, *La furia de las imágenes*, 32.

30. I have addressed how certain artists have dealt with the sensory constraints of imagining animal worlds: Concepción Cortés Zulueta, "How Does a Snail See the World? Imagining Non-human Animals' Visual *Umwelten*," in *Meta- and Inter-images in Contemporary Visual Art and Culture*, ed. Carla Taban, 263–79 (Louvain, Belgium: Leuven University Press, 2013).

31. Rosemary-Claire Collard has discussed the ambivalence of films, and particularly of wildlife films, since they have the potential to either animate or enliven animals and viewers or reinforce violent hierarchies; Rosemary-Claire Collard, "Electric Elephants and the Lively/Lethal Energies of Wildlife Documentary Film: Electric Elephants and Wildlife Documentary Film," *Area* 48, no. 4 (December 2016): 472–79.

32. Chekhov's short story is titled "Gusev" (1890), and the full quote, according to the translation stored by Project Gutenberg, is: "After that another dark body appeared. It was a shark. It swam under Gusev with dignity and no show of interest, as though it did not notice him, and sank down upon its back, then it turned belly upwards, basking in the warm, transparent water and languidly opened its jaws with two rows of teeth. The harbour pilots are delighted, they stop to see what will come next. After playing a little with the body the shark nonchalantly puts its jaws under it, cautiously touches it with its teeth, and the sailcloth is rent its full length from head to foot; one of the weights falls out and frightens the harbour pilots, and striking the shark on the ribs goes rapidly to the bottom." Anton Chekov, "Gusev," in *The Witch and Other Stories*, prod. James Rusk and David Widger, Project Gutenberg, accessed November 1, 2018, http://www.gutenberg.org/files/1944/1944-h/1944-h.htm.

33. Gregg Mitman, *Reel Nature: America's Romance with Wildlife on Film* (Seattle: University of Washington Press, 2009), 59.

34. I won't go into detail here, but elsewhere I have argued that Alexander Pschera's book *Animal Internet* is an exceedingly optimistic instance of this aspiration to an intimate and technological kind of transparency, while the interactive web documentary *Bear 71* shows its dark sides, unbalanced hierarchies and surveillance associations; Cortés Zulueta, "That Seagull Stole My Camera (and My Shot)!"

35. Donna Haraway, *When Species Meet* (Minneapolis: University of Minnesota Press, 2008), 252.

36. "Lions—Spy in the Den," John Downer Productions, accessed August 10, 2017, http://jdp.co.uk/programmes/lions-spy-in-the-den.

37. Actually, "Could animals be more like us than we ever believed possible?" is the sentence that concludes the introduction to the five episodes of the *Spy in the Wild* series.

38. JohnDownerProd, *Interview with John Downer: POV-Unique Perspectives & Immersive Viewpoints*, 2015, https://www.youtube.com/watch?v=2yZpfKLceNI.

39. For more examples and a more detailed discussion of the *Spy in the Wild* series, see Cortés Zulueta, "That Seagull Stole My Camera (and My Shot)!"

40. This is exemplified by conventional dioramas, so paradigmatic of some Natural History Museums.

41. In fact, I have borrowed the term "bubbles" from Uexküll's writings; Jakob von Uexküll and Georg Kriszat, "A Stroll through the Worlds of Animals and Men: A Picture Book of Invisible Worlds," in *Instinctive Behavior: The Development of a Modern Concept*, ed. Claire H. Schiller, 5–80 (New York: International University Press, 1957).

42. Mitman, *Reel Nature*, 84. I briefly addressed this shift, and the related recent developments and their aesthetics in Cortés Zulueta, "That Seagull Stole My Camera (and my Shot)!"

43. JohnDownerProd, *Interview with John Downer*.

44. Anne Friedberg, *The Virtual Window: From Alberti to Microsoft* (Cambridge, MA: MIT Press, 2009).

EPILOGUE: ENTANGLEMENTS AND VISIBILITIES

1. James Scott, *Seeing Like a State: How Certain Schemes to Improve the Human Condition Have Failed* (New Haven, CT: Yale University Press, 1999); John Berger, *About Looking* (New York: Pantheon Books, 1980).

2. Haraway acknowledges the work of her colleague Karen Barad, whose book *Meeting the Universe Halfway: Quantum Physics and the Entanglement of Matter and Meaning* (Durham, NC: Duke University Press, 2006), influenced her own word choice.

3. Hunting as play calls to mind Yi Fu Tuan's focus on the use of the words "game" and "sport" by hunters in *Dominance and Affection: The Making of Pets* (New Haven, CT: Yale University Press, 1985).

4. John Berger, *Ways of Seeing* (New York: Penguin Books, 1972).

5. Matthew Edney, *Mapping an Empire: The Geographical Construction of Colonial India 1765–1843* (Chicago: University of Chicago Press, 1997).

6. Harriet Ritvo has written extensively on the classifying imagination, but Geoffrey Bowker and Susan Leigh Starr use this passage to highlight the compromises and omissions that are baked into any system that tries to quantify previously invisible or unknown information. Bowker and Starr, *Sorting Things Out: Classification and Its Consequences* (Cambridge, MA: MIT Press, 2000), 53–55.

SELECTED BIBLIOGRAPHY

The following bibliography contains a selection of significant works cited in the contributions.

Abberley, Will, ed. *Underwater Worlds: Submerged Visions in Science and Culture*. Newcastle upon Tyne, UK: Cambridge Scholars, 2018.

Adams, Catherine, and Terrie Lynn Thompson. *Researching a Posthuman World: Interviews with Digital Objects*. London: Palgrave Macmillan, 2016.

Adams, Ross Exo. "Mare Magnum: Urbanization of the Sea." In *Association of American Geographers Conference 2015* (Chicago: AAG, 2015).

Aitken, Gill. "Conservation and Individual Worth." *Environmental Values* 6 (1997): 439–54.

Alagona, Peter S. "Introduction: Fifty Years of Wildlife in America." *Environmental History* 16, no. 3 (2011): 391–97.

Alexander, Jennifer Karns. *The Mantra of Efficiency: From Waterwheel to Social Control*. Baltimore: Johns Hopkins University Press, 2008.

Anderson, Julie, and Neil Pemberton. "Walking Alone: Aiding the War and Civilian Blind in the Inter-war Period." *European Review of History—Revue Européenne d'Histoire* 14, no. 4 (2007): 459–79.

Anderson, Virginia DeJohn. *Creatures of Empire: How Domestic Animals Transformed Early America*. Oxford: Oxford University Press, 2006.

Angelo, Hillary, and David Wachsmuth. "Urbanizing Urban Political Ecology: A Critique of Methodological Cityism." *International Journal of Urban and Regional Research* 39, no. 1 (2015): 16–27.

Anker, Peder. *From Bauhaus to Ecohouse: A History of Ecological Design.* Baton Rouge: LSU Press, 2010.

Antonello, Alessandro. "Life, Ice and Ocean: Contemporary Antarctic Spaces." In *Handbook on the Politics of Antarctica,* edited by Klaus Dodds, Alan D. Hemmings, and Peder Roberts, 167–82. Cheltenham, UK: Edward Elgar, 2017.

Atwood, Christopher P. *Encyclopedia of Mongolia and the Mongol Empire.* New York: Facts on File, 2004.

Baar, Monika. "Prosthesis for the Body and for the Soul: The Origins of Guide Dog Provision for Blind Veterans in Interwar Germany." *First World War Studies* 6, no. 1 (2015): 81–98.

Barad, Karen. *Meeting the Universe Halfway: Quantum Physics and the Entanglement of Matter and Meaning.* Durham, NC: Duke University Press, 2006.

Barad, Karen. "Posthumanist Performativity: Toward an Understanding of How Matter Comes to Matter." *Signs: Journal of Women in Culture & Society* 28, no. 3 (2003): 801–31.

Bateson, Gregory. *Steps to an Ecology of Mind.* Chicago: University of Chicago Press, 2000.

Bauer, Wolfgang. *Tsingtao 1914 bis 1931: Japanische Herrschaft, wirtschaftliche Entwicklung und die Rückkehr der deutschen Kaufleute.* Munich: Iudicium, 1999.

Beach, Hugh, and Florian Stammler. "Human-Animals Relations in Pastoralism." *Nomadic Peoples* 10, no. 2 (2006): 6–30.

Bellacasa, María Puig de la. *Matters of Care: Speculative Ethics in More than Human Worlds.* Minneapolis: University of Minnesota Press, 2017.

Bennett, Jane. *Vibrant Matter: A Political Ecology of Things.* Durham, NC: Duke University Press, 2010.

Benson, Etienne. *Wired Wilderness: Technologies of Tracking and the Making of Modern Wildlife.* Baltimore: Johns Hopkins University Press, 2010.

Berger, John. *Ways of Seeing.* New York: Penguin Books, 1972.

Berger, John. "Why Look at Animals?" In *Animals and Society: Critical Concepts in the Social Sciences,* edited by Rhoda Wilkie and David Inglis, 64–78. London: Routledge, 2007.

Berkes, Fikret. *Sacred Ecology.* New York: Routledge, 1999.

Bernard, Russel H., and Pertii J. Pelto. "Technology and Social Change: Introduction." In *Technology and Social Change,* edited by H. Russell Bernard and Pertii J. Pelto, 1–10. Prospect Height, IL: Waveland Press, 1987 [1972].

Birke, Lynda, Mette Bryld, and Nina Lykke, "Animal Performances: An Exploration of Intersections between Feminist Science Studies and Studies of Human/Animal Relationships." *Feminist Theory* 5, no. 2 (2004): 167–83.

Black, Megan. *The Global Interior: Mineral Frontiers and American Power.* Cambridge, MA: Harvard University Press, 2019.

Bodenhamer, David J., John Corrigan, and Trevor M. Harris. *Deep Maps and Spatial Narratives.* Bloomington: Indiana University Press, 2015.

Bourke, Joanna. *Deep Violence: Military Violence, War Play, and the Social Life of Weapons*. Berkeley, CA: Counterpoint Press, 2015.

Bourke, Joanna. *An Intimate History of Killing: Face-to-Face Killing in 20th Century Warfare*. New York: Basic Books, 1999.

Bowker Geoffrey C., and Susan Leigh Starr. *Sorting Things Out: Classification and Its Consequences*. Cambridge, MA: MIT Press, 2000.

Brantz, Dorothee. "The Domestication of Empire: Human-Animal Relations at the Intersection of Civilization, Evolution, and Acclimatization." In *A Cultural History of Animals in the Age of Empire*, edited by Kathleen Kete, 73–94. Oxford: Berg, 2011.

Brenner, Neil. "Urban Theory without an Outside." *Harvard Design Magazine*, no. 37 (2014): 42–47.

Brenner, Neil, and Christian Schmid. "Planetary Urbanisation." In *Urban Constellations*, edited by Matthew Gandy, 10–13. Berlin: Jovis Verlag, 2011.

Brenner, Neil, and Nik Theodore. "Cities and the Geographies of Actually Existing Neoliberalism." In *Critique of Urbanization: Selected Essays*, edited by Neil Brenner, 42–68. Basel, Switzerland: Birkhäuser, 2017 [2002].

Brown, Bill. "Thing Theory." *Critical Inquiry* 28, no. 1 (2001): 1–22.

Burnett, D. Graham. *The Sounding of the Whale: Science & Cetaceans in the Twentieth Century*. Chicago: University of Chicago Press, 2013.

Butler, Judith. *Gender Trouble: Feminism and the Subversion of Identity*. London: Routledge, 1990.

Callon, Michel. "Some Elements of a Sociology of Translation: Domestication of the Scallops and the Fishermen of St Brieuc Bay." In *Power, Action and Belief: A New Sociology of Knowledge?*, edited by John Law, 196–233. London: Routledge & Kegan Paul, 1986.

Campbell, Colleen, and Tina Loo. 2022. "Making Tracks: A Grizzly and Entangled History." In *Traces of the Animal Past: Methodological Challenges in Animal History*, edited by Jennifer Bonnell and Sean Kheraj. Calgary, AB: University of Calgary Press, 2022.

Campbell, Douglas Ian, and Patrick Michael Whittle. *Resurrecting Extinct Species: Ethics and Authenticity*. Cham, Switzerland: Palgrave Macmillan, 2017.

Cannon, Walter B. *Bodily Changes in Pain Hunger Fear and Rage*. New York: D. Appleton, 1927.

Cannon, Walter B. "The James-Lange Theory of Emotions: A Critical Examination and an Alternative Theory." *American Journal of Psychology* 100, no. 3–4 (Fall-Winter 1987): 567–86.

Carbone, Larry. *What Animals Want: Expertise and Advocacy in Laboratory Animal Welfare Policy*. Oxford: Oxford University Press, 2004.

Carr, Nicholas. *The Shallows: What the Internet Is Doing to Our Brains*. New York: W. W. Norton, 2010.

Carruthers, Douglas, and Jack Miller Humphrey. *Unknown Mongolia: A Record of Travel and Exploration in North-West Mongolia and Dzungaria*. London: Hutchinson, 1914.

Cartmill, Matt. *A View to a Death in the Morning: Hunting and Nature through History*. Cambridge, MA: Harvard University Press, 1996.

Ceruzzi, Paul E. *GPS*. Cambridge, MA: MIT Press, 2018.

Christ, Robert D., and Robert L. Wernli Sr. *The ROV Manual: A User Guide for Remotely Operated Vehicles*. Oxford: Elsevier Science, 2013.

Cincinnati, Noah. "Too Sullen for Survival: Historicizing Gorilla Extinction, 1900–1930." In *The Historical Animal*, edited by Susan Nance, 166–83. Syracuse, NY: Syracuse University Press, 2015.

Cohen, Benjamin R. *Pure Adulteration: Cheating on Nature in the Age of Manufactured Food*. Chicago: University of Chicago Press, 2019.

Cohen, Margaret, and Killian Quigley. *The Aesthetics of the Undersea*. Abingdon, UK: Routledge, 2019.

Coleman, John T. *Vicious: Wolves and Men in America*. New Haven: Yale University Press, 2004.

Conrad, Sebastian. *German Colonialism: A Short History*. Cambridge: Cambridge University Press, 2012.

Conradson, David. "Geographies of Care: Spaces, Practices, Experiences." *Social & Cultural Geography* 4 (2003): 451–54.

Cortés Zulueta, Concepción. "How Does a Snail See the World? Imagining Nonhuman Animals' Visual *Umwelten*." In *Meta- and Inter-Images in Contemporary Visual Art and Culture*, edited by Carla Taban, 263–79. Louvain, Belgium: Leuven University Press, 2013.

Cortés Zulueta, Concepción. "That Seagull Stole My Camera (and My Shot)! Overlapping Metaphorical and Physical Distances in the Human-Animal-Camera Triad." In *Critical Distance in Documentary Media*, edited by Gerda Cammaer, Blake Fitzpatrick, and Bruno Lessard, 231–56. Cham, Switzerland: Palgrave Macmillan, 2018.

Crampton, Jeremy W., and Stuart Elden, eds. *Space, Knowledge, and Power: Foucault and Geography*. Aldershot, UK: Ashgate, 2007.

Crary, Jonathan. *Techniques of the Observer: On Vision and Modernity in the Nineteenth Century*. Cambridge, MA: MIT Press, 1990.

Cronon, William. *Nature's Metropolis: Chicago and the Great West*. New York: W. W. Norton, 1995.

Cronon, William. "The Trouble with Wilderness; or, Getting Back to the Wrong Nature." In *Uncommon Ground: Rethinking the Human Place in Nature*, edited by William Cronon, 69–90. New York: W. W. Norton, 1995.

Crylen, Jonathan Christopher. "The Cinematic Aquarium: A History of Undersea Film." PhD diss., University of Iowa, 2015.

Danell, Kjell, et al., eds. *Jaktens historia i Sverige: Vilt–människa–samhälle–kultur*. Stockholm: Liber, 2016.

Davis, Susan G. *Spectacular Nature: Corporate Culture and the Sea World Experience*. Berkeley: University of California Press, 1997.

Dean, Joanna, Darcy Ingram, and Christabelle Sethna, eds. *Animal Metropolis: Histories of Human-Animal Relations in Urban Canada*. Calgary, AB: University of Calgary Press, 2017.

Demuth, Bathsheba. *Floating Coast: An Environmental History of the Bering Strait*. New York: W.W. Norton, 2019.

Derickson, Kate D. "Urban Geography I: Locating Urban Theory in the 'Urban Age.'" *Progress in Human Geography* 39, no. 5 (2015): 647–57.

Derry, Margaret. *Masterminding Nature: The Breeding of Animals, 1750–2010*. Toronto: University of Toronto Press, 2015.

Deshen, Shlomo, and Hilda Deshen, "On Social Aspects of the Usage of the Usage of Guide Dogs and Long-Canes." *Sociological Review* 37, no. 1 (1989): 89–103.

Dirke, Karin. "Changing Narratives of Human-Large Carnivore Encounters." In *Shared Lives of Humans and Animals: Animal Agency in the Global North*, edited by Taina Syrjämaa and Tuomas Räsenen, 163–78. London: Routledge, 2017.

Dirke, Karin. *De Värnlösas Vänner*. Stockholm: Almqvist & Wiksell, 2000.

Doordan, Dennis. "Simulated Seas: Exhibition Design in Contemporary Aquariums." *Design Issues* 11, no. 2 (Summer 1995): 3–10.

Edney, Matthew. *Mapping an Empire: The Geographical Construction of Colonial India 1765–1843*. Chicago: University of Chicago Press, 1997.

Eide, Øyvind. *Media Boundaries and Conceptual Modelling: Between Texts and Maps*. Basingstoke, UK: Palgrave Macmillan, 2015.

Elias, Ann. *Coral Empire: Underwater Oceans, Colonial Tropics, Visual Modernity*. Durham, NC: Duke University Press, 2019.

Emel, Jody, and Jennifer Wolch. "Witnessing the Animal Moment." In *Animal Geographies: Place, Politics, and Identity in the Nature-Culture Borderlands*, edited by Jody Emel and Jennifer Wolch, 1–23. London: Verso, 1998.

Fangjun, Cao. "Modernization Theory and China's Road to Modernization." *Chinese Studies in History* 43, no. 1 (2009): 7–16.

Farnham, Timothy J. *Saving Nature's Legacy: Origins of the Idea of Biodiversity*. New Haven, CT: Yale University Press, 2007.

Ferret, Carole. "La figure atemporelle du 'nomade des steppes.'" In *La préhistoire des autres: Perspectives archéologiques et anthropologiques*, edited by Nathan Schlanger and Anne-Christine Taylor, 167–82. Paris: La Découverte, 2012.

Fijn, Natasha. *Living with Herds: Human-Animal Coexistence in Mongolia*. Cambridge: Cambridge University Press, 2011.

Fitzgerald, Deborah Kay. *Every Farm a Factory: The Industrial Ideal in American Agriculture*. New Haven, CT: Yale University Press, 2003.

Flack, Andrew. *The Wild Within: Histories of a Landmark British Zoo*. Charlottesville: University of Virginia Press, 2018.

Fontcuberta, Joan. *La furia de las imágenes: Notas sobre la postfotografía*. Barcelona: Galaxia Gutenberg, 2016.

Franklin, Alex, and Nora Schuurman. "Aging Animal Bodies: Horse Retirement Yards as Relational Spaces of Liminality, Dwelling and Negotiation." *Social & Cultural Geography* 20 (2019): 918–37.

Fraser, Richard. "Motorcycles on the Steppe: Skill, Social Change, and New Technologies in Postsocialist Northern Mongolia." *Nomadic Peoples* 22, no. 2 (2018): 330–68.

Frederiksen, Tomas, and Matthew Himley. "Tactics of Dispossession: Access, Power, and Subjectivity at the Extractive Frontier." *Transactions of the Institute of British Geographers* 45, no. 1 (2020): 50–64.

Friedberg, Anne. *The Virtual Window: From Alberti to Microsoft*. Cambridge, MA: MIT Press, 2009.

Galis, Vasilis. "Enacting Disability: How Can Science and Technology Studies Inform Disability Studies?" *Disability & Society* 26, no. 7 (2011): 825–38.

Gaudiosi, Massimiliano. "Marey's Aquarium: The Underwater World and the Archaeology of Cinema." In *Underwater Worlds: Submerged Visions in Science and Culture*, edited by Will Abberley, 168–88. Newcastle upon Tyne, UK: Cambridge Scholars, 2018.

Gibson, James. *Ecological Approach to Visual Perception*. New York: Taylor & Francis Group, 1986.

Goodman, David S. G. "The Campaign to 'Open Up the West': National, Provincial-Level and Local Perspectives." *China Quarterly* 178 (2004): 317–34.

Greene, Ann Norton. *Horses at Work: Harnessing Power in Industrial America*. Cambridge, MA: Harvard University Press, 2008.

Gruen, Lori, and Kari Weil. "Teaching Difference: Sex, Gender, Species." In *Teaching the Animal: Human–Animal Studies across the Disciplines*, edited by Margo DeMello, 127–42. New York: Lantern Books, 2010.

Gyokuji, Ran. *Chintao no toshi keisei shi, 1897–1945: Shijo keizai no keisei to tenkai*. Kyoto: Shibunkaku, 2009.

Hamera, Judith. *Parlor Ponds: The Cultural Work of the American Home Aquarium*. Ann Arbor: University of Michigan Press, 2012.

Haraway, Donna. "Anthropocene, Capitalocene, Plantationocene, Chthulucene: Making Kin." *Environmental Humanities* 6 (2015): 159–65.

Haraway, Donna J. *Primate Visions: Gender, Race, and Nature in the World of Modern Science*. New York: Routledge, 1989.

Haraway, Donna J. *Simians, Cyborgs, and Women: The Reinvention of Nature*. New York: Routledge, 1988.

Haraway, Donna. *Staying with the Trouble: Making Kin in the Cthulhucene*. Durham, NC: Duke University Press, 2016.

Haraway, Donna. *When Species Meet*. Minneapolis: University of Minnesota Press, 2008.

Hare, Brian, and Michael Tomasello, "Human-like Social Skills in Dogs?" *Trends in Cognitive Sciences* 9, no. 9 (2005): 439–44.

Hayward, Eva. "Sensational Jellyfish: Aquarium Affects and the Matter of Immersion." *differences: A Journal of Feminist Cultural Studies* 25, no. 3 (2012): 161–96.

Hemmings, Alan D., Klaus Dodds, and Peder Roberts. "Introduction: The Politics of Antarctica." In *Handbook on the Politics of Antarctica*, edited by Klaus Dodds, Alan D. Hemmings, and Peder Roberts, 1–20 (Cheltenham, UK: Edward Elgar, 2017).

Hersh, Marion A., and Michael A. Johnson. "Mobility: An Overview." In *Assistive Technology for Visually Impaired and Blind People*, edited by Marion A. Hersh and Michael A. Johnson, 167–208. London: Springer, 2008.

Hodgetts, Timothy, and Jamie Lorimer. "Animals' Mobilities." *Progress in Human Geography* 44, no. 1. (2020): 4–26.

Huang, Brian, et al., "Take Only Pictures, Leave Only . . . Fear? The Effects of Photography on the West Indian Anole Anolis Cristatellus." *Current Zoology* 57, no. 1 (2011): 77–82.

Humphrey, Caroline, and David Sneath. *The End of Nomadism? Society, State and the Environment in Inner Asia*. Durham, NC: Duke University Press, 1999.

Inglis, Julia. *Traditional Ecological Knowledge: Concepts and Cases*. Ottawa: International Program on Traditional Ecological Knowledge, International Development Research Centre, 1993.

Iselin, Lilian. "Translocal Practices on the Tibetan Plateau: Motorised Mobility of Pastoralists and Spatial Transformations." *Nomadic Peoples* 18 (2014): 15–37.

Jacoby, Karl. *Crimes against Nature: Squatters, Poachers, Thieves, and the Hidden History of American Conservation*. Berkeley: University of California Press, 2001.

Jones, Henry. "Lines in the Ocean: Thinking with the Sea about Territory and International Law." *London Review of International Law* 4, no. 2 (2016): 307–43.

Jones, Karen R. *Wolf Mountains: A History of Wolves along the Great Divide*. Calgary, AB: University of Calgary Press, 2002.

Jørgensen, Dolly. "Artifacts and Habitats." In *The Routledge Companion to Environmental Humanities*, edited by Ursula K. Heise, Jon Christensen, and Michelle Niemann, 138–43. London: Routledge 2017.

Jørgensen, Dolly. "Mixing Oil and Water: Naturalizing Offshore Oil Platforms in American Aquariums." In *Oil Culture*, edited by D. Worden and R. Barrett, 267–88. Minneapolis: University of Minnesota Press, 2014.

Jørgensen, Dolly. "Not by Human Hands: Five Technological Tenets for Environmental History in the Anthropocene." *Environment & History* (2014): 479–89.

Jørgensen, Dolly. *Recovering Lost Species in the Modern Age: Histories of Longing and Belonging*. Cambridge, MA: MIT Press, 2019.

Jørgensen, Dolly, Finn Arne Jørgensen, and Sara B. Pritchard, eds. *New Natures: Joining Environmental History with Science and Technology Studies*. Pittsburgh: University of Pittsburgh Press, 2013.

Jütte, Robert. *A History of the Senses: From Antiquity to Cyberspace*. Translated by J. Lynn. Cambridge: Polity, 2005.

Khazanov, Anatoly. *Nomads and the Outside World*. Cambridge: Cambridge University Press, 1983.

Knezevic, Irena. "Hunting and Environmentalism: Conflict or Misperceptions." *Human Dimensions of Wildlife* 14, no. 1 (2009): 12–20.

Koolmees, Peter. "Veterinary Inspection and Food Hygiene in the Twentieth Century." In *Food, Science, Policy and Regulation in the Twentieth Century*, edited by David F. Smith and Jim Phillips, 63–78. London: Routledge, 2000.

Koski, Leena, and Pia Bäcklund. "Whose Agency? Humans and Dogs in Training." In *Shared Lives of Humans and Animals: Animal Agency in the Global North*, edited by Tuomas Räsänen and Taina Syrjämaa, 11–23. London: Routledge, 2017.

Kranzberg, Melvin. "Technology and History: 'Kranzberg's Laws.'" *Technology and Culture*, 27, no. 3 (1986): 544–60.

Kreeger, Terry J. "The Internal Wolf: Physiology, Pathology, and Pharmacology." In *Wolves: Behavior, Ecology and Conservation*, edited by David Mech and Luigi Boitani, 192–217. Chicago: University of Chicago Press, 2006.

Kroglund, Torolf E. *Jegerhjerte: Jakten på jegerens hemmelighet*. Oslo: Pax, 2015.

Lachapelle, Sofie, and Heena Mistry. "From the Waters of the Empire to the Tanks of Paris: The Creation and Early Years of the Aquarium Tropical, Palais de la Porte Dorée." *Journal of the History of Biology* 47 (2014): 1–27.

Lähdesmäki, Heta. *Susien paikat: Ihminen ja susi 1900-luvun Suomessa*. Nykykulttuurin tutkimuskeskuksen julkaisuja 127. Jyväskylä, Finland: University of Jyväskylä, 2020.

Latour, Bruno. *The Pasteurization of France*. Translated by Alan Sheridan. Cambridge, MA: Harvard University Press, 1993.

Latour, Bruno. *Politics of Nature: How to Bring the Sciences into Democracy*. Cambridge, MA: Harvard University Press, 2004.

Latour, Bruno. *Reassembling the Social: An Introduction to Actor-Network Theory*. Oxford: Oxford University Press, 2005.

Latour, Bruno. *We Have Never Been Modern*. Translated by Catherine Porter. Cambridge, MA: Harvard University Press, 1993.

Lavrillier, Alexandra. "Nomadisme et adaptations sédentaires chez les Évenks de Sibérie postsoviétique: 'Jouer' pour vivre avec et sans chamanes." PhD diss., École Pratique des Hautes Études, 2005.

Leach, William R. *Land of Desire: Merchants, Power, and the Rise of a New American Culture*. New York: Vintage Books, 2011.

Lederer, Susan E. *Subjected to Science: Human Experimentation in America before the Second World War*. Baltimore: Johns Hopkins University Press, 1997.

Lee, Paula Young, ed. *Meat, Modernity, and the Rise of the Slaughterhouse*. Durham: University of New Hampshire Press, 2008.

Lefebvre, Henri. *The Production of Space*. London: Wiley-Blackwell, 1991.

Lefebvre, Henri. "Right to the City." In *Writings on Cities*, edited by Eleonore Kofman and Elizabeth Lebas, 63–184. Oxford: Blackwell, 1996 [1968].

Lefebvre, Henri. *The Urban Revolution*. Minneapolis: University of Minnesota Press, 2003 [1970].

Le Gall, Guillaume. "Dioramas aquatiques: Théophile Gautier visite l'aquarium du Jardin d'acclimatation." *Culture & Musées* 32 (2018): 81–106.

Lehman, Jessica S. "Volumes beyond Volumetrics: A Response to Simon Dalby's 'The Geopolitics of Climate Change.'" *Political Geography* 37 (2013): 51–52.

Leutner, Mechthild, and Klaus Mühlhahn. *"Musterkolonie Kiautschou": Die Expansion des Deutschen Reiches in China: Deutsch-chinesische Beziehungen 1897 bis 1914. Eine Quellensammlung*. Berlin: Akademie Verlag, 1997.

Linnell, John D. C., John Odden, Martin E. Smith, Ronny Aanes, and Jon E. Swenson. "Large Carnivores That Kill Livestock: Do 'Problem Individuals' Really Exist?" *Wildlife Society Bulletin* 27, no. 3 (1999): 698–705.

Loo, Tina. *States of Nature: Conserving Canada's Wildlife in the Twentieth Century*. Vancouver: UBC Press, 2006.

Magnus, Riin. "The Semiotic Challenges of Guide Dog Teams: The Experiences of German, Estonian and Swedish Guide Dog Users." *Biosemiotics* 9, no. 2 (2015): 267–85.

Maier, Jonathan R. A., and Georges M. Fadel. "An Affordance-Based Approach to Architectural Theory, Design, and Practice." *Design Studies* 30, no. 4 (2009): 393–414.

Marois, Alexandra. "D'un habitat mobile à un habitat fixe: Fondements et changements de l'orientation dans l'espace domestique mongol." *Études Mongoles et Sibériennes, Centrasiatiques et Tibétaines* 36–37 (2006): 207–37.

Massey, Doreen. *For Space*. London: Sage, 2005.

Masson, Jeffrey Moussaieff, and Susan McCarthy. *When Elephants Weep: The Emotional Lives of Animals*. London: Vintage, 1995.

Mech, David, and Luigi Boitani. "Preface." In *Wolves: Behavior, Ecology and Conservation*, edited by David Mech and Luigi Boitani, xi–xii. Chicago: University of Chicago Press, 2006.

Merrifield, Andy. "The Urban Question under Planetary Urbanisation." In *Implosions/Explosions: Towards a Study of Planetary Urbanization*, edited by Neil Brenner, 164–80. Berlin: Jovis, 2014.

Merritt, Michele. *Minding Dogs: Humans, Canine Companions, and a New Philosophy of Cognitive Science*. Athens: University of Georgia Press, 2021.

Milam, Erika Lorraine. *Creatures of Cain: The Hunt for Human Nature in Cold War America*. Princeton, NJ: Princeton University Press, 2018.

Miller, Ian Jared. *The Nature of Beasts: Empire and Exhibition at the Tokyo Imperial Zoo*. Oakland: University of California Press, 2021.

Milligan, Christine, and Janine Wiles. "Landscape of Care." *Progress in Human Geography* 34 (2010): 736–54.

Mitman, Gregg. *Reel Nature: America's Romance with Wildlife on Film*. Madison: University of Wisconsin Press, 2009.

Mitman, Gregg. *The State of Nature: Ecology, Community, and American Social Thought, 1900–1950*. Chicago: University of Chicago Press, 1992.

Moser, Ingunn. "Disability and the Promises of Technology: Technology, Subjectivity and Embodiment within an Order of the Normal." *Information, Communication & Society* 9 (2006): 373–95.

Muka, Samantha. *Oceans under Glass: Tank Craft and the Sciences of the Sea*. Chicago: University of Chicago Press, 2023.

Murdoch, Jon. *Post-structuralist Geography: A Guide to Relational Space*. London: Sage, 2006.

Naderi, Szima, Adam Miklósi, Antal Dóka, and Vilmos Csányi. "Co-operative Interactions between Blind Persons and Their Dogs." *Applied Animal Behaviour Science* 74, no. 1 (2001): 59–80.

Nance, Susan. "Introduction." In *The Historical Animal*, edited by Susan Nance. Syracuse, NY: Syracuse University Press, 2015.

Nash, Roderick Frazier. *The Rights of Nature: A History of Environmental Ethics*. Madison: University of Wisconsin Press, 1989.

Nestle, Marion, and Malden Nesheim. *Why Calories Count: From Science to Politics*. Berkeley: University of California Press, 2012.

Nordenram, Åke. *Svenska jaktens historia: Från forntid till nutid*. Örkelljunga, Sweden: Bokförlaget Settern, 2001.

Norman, Don. *The Design of Everyday Things*. New York: Basic Books, 2013.

Nyrén, Ulf. *Rätt till jakt: En studie av den svenska jakträtten ca 1600–1789*. PhD diss., Göteborgs Universitet, 2012.

Panelli, Ruth. "More-Than-Human Social Geographies: Posthuman and Other Possibilities." *Progress in Human Geography* 34, no. 1 (2010): 79–87.

Pawley, Emily. "Accounting with the Fields: Chemistry and Value in Nutriment in American Agricultural Improvement, 1835–1860." *Science as Culture* 19, no. 4 (2010): 461–82.

Pawley, Emily. *The Nature of the Future: Agriculture, Science, and Capitalism in the Antebellum North*. Chicago: University of Chicago Press, 2020.

Perren, Richard. *Taste, Trade and Technology: The Development of the International Meat Industry since 1840*. Aldershot, UK: Ashgate, 2006.

Philo, Chris. "(In)secure Environments and the Domination of Nature." *Geographical Journal* 181 (2015): 322–37.

Philo, Chris, and Chris Wilbert. *Animal Spaces, Beastly Places: New Geographies of Human-Animal Relations*. London: Routledge, 2000.

Prott, Lyndel V. "Hunting as Intangible Heritage: Some Notes on Its Manifestations." *International Journal of Cultural Property* 14, no. 3 (2007): 385–98.

Pschera, Alexander. *Animal Internet: Nature and the Digital Revolution*. Translated by Elisabeth Lauffer. New York: New Vessel Press, 2014.

Pulliainen, Erkki, and Lassi Rautiainen. *Suurpetomme: Karhu, susi, ahma, ilves*. Kotka, Finland: Articmedia, 1999.

Rader, Karen A., and Victorian E. M. Cain. *Life on Display: Revolutionizing U.S. Museums of Science and Natural History in the Twentieth Century*. Chicago: University of Chicago Press, 2014.

Radin, Joanna. *Life on Ice: A History of New Uses for Cold Blood*. Chicago: University of Chicago Press, 2017.

Räsänen, Tuomas. "Converging Environmental Knowledge: Re-evaluating the Birth of Modern Environmentalism in Finland." *Environment and History* 18 (2012): 159–81.

Räsänen, Tuomas. "Does a Dead Wild Animal Have Agency? The White-Tailed Eagle as a Catalyst for an Ideational Revolution in Finland." In *Shared Lives of Humans and Animals: Animal Agency in the Global North*, edited by Tuomas Räsänen and Taina Syrjämaa, 93–104. London: Routledge, 2017.

Rickards, Lauren. "Metaphor and the Anthropocene: Presenting Humans as a Geological Force." *Geographical Research* 53, no. 3 (2015): 280–87.

Rickards, Lauren, Brendan Gleeson, Mark Boyle, and Cian O'Callaghan. "Urban Studies after the Age of the City." *Urban Studies* 53, no. 8 (2016): 1523–41.

Roberts, Peder, and Dolly Jørgensen. "Animals as Instruments of Norwegian Imperial Authority in the Interwar Arctic." *Journal for the History of Environment and Society* 1, no. 1 (2016): 65–87.

Rockström, Johan, W. Steffen, K. Noone, Å. Persson, F. S. Chapin III, E. Lambin, T. M. Lenton, et al. "Planetary Boundaries: Exploring the Safe Operating Space for Humanity." *Ecology and Society* 14, no. 2 (2009): n.p.

Rose, Lisle Abbott. *Explorer: The Life of Richard E. Byrd*. Columbia: University of Missouri Press, 2008.

Rozwadowski, Helen M. "'Bringing Humanity Full Circle Back into the Sea': *Homo aquaticus*, Evolution, and the Ocean." *Environmental Humanities* 14, no. 1 (2022): 1–28.

Russell, Edmund. *Greyhound Nation: A Coevolutionary History of England, 1200–1900*. Cambridge: Cambridge University Press, 2018.

Rutherford, Stephanie. "A Resounding Success? Howling as a Source of Environmental History." In *Methodological Challenges in Nature-Culture and Environmental History Research*, edited by Jocelyn Thorpe, Stephanie Rutherford, and L. Anders Sandberg, 43–47. New York: Routledge, 2017.

Rutherford, Stephanie. *Villain, Vermin, Icon, Kin. Wolves and the Making of Canada*. Montreal: McGill-Queen's University Press, 2022.

Saraiva, Tiago. *Fascist Pigs: Technoscientific Organisms and the History of Fascism*. Cambridge, MA: MIT Press, 2016.

Schrepfer, Susan R., and Phil Scranton, eds. *Industrializing Organisms: Introducing Evolutionary History*. New York: Routledge, 2004.

Schrijver, Nico. *Sovereignty over Natural Resources: Balancing Rights and Duties*. Cambridge: Cambridge University Press, 1997.

Shepherd, Paul. *Man in the Landscape: A Historic View of the Esthetics of Nature*. Athens: University of Georgia Press, 2002.

Shick, J. Malcolm. "Toward an Aesthetic Marine Biology." *Art Journal* 67, no. 4 (2008): 62–86.

Shinji, Asada. *Doitsu tochika no Chintao: Keizaiteki jiyushugi to shokuminchi shakai chitsujo*. Tokyo: University of Tokyo Press, 2011.

Sinclair, Anthony R. E., John M. Fryxell, and Graeme Gaughley. *Wildlife Ecology, Conservation and Management*. Chichester, UK: Wiley Blackwell, 2014.

Smith-Howard, Kendra. *Pure and Modern Milk: An Environmental History since 1900*. Oxford : Oxford University Press, 2013.

Sneath, David. *Changing Inner Mongolia: Pastoral Mongolian Society and the Chinese State*. Oxford: Oxford University Press, 2000.

Sneath, David. "Notions of Rights over Land and the History of Mongolian Pastoralism." *Inner Asia* 3, no. 1 (2001): 41–58.

Snively, Gloria, and John Corsiglia. "Indigenous Science: Proven, Practical and Timeless." In *Knowing Home: Braiding Indigenous Science with Western Science*, Book 1, edited by Gloria Snively and Wanosts'a 7 Lorna Williams. Victoria, BC: University of Victoria Press, 2016.

Söderqvist, Thomas. "The Ecologists: From Merry Naturalists to Saviours of the Nation: A Sociologically Informed Narrative Survey of the Ecologization of Sweden 1895–1975." PhD diss., Göteborgs Universitet, 1986.

Soja, Edward. *Thirdspace: Journeys to Los Angeles and Other Real-and-Imagined Places*. London: Wiley-Blackwell, 1996.

Sontag, Susan. *On Photography*. New York: Farrar, Straus and Giroux, 1977.

Specht, Joshua. *Red Meat Republic: A Hoof-to-Table History of How Beef Changed America*. Princeton, NJ: Princeton University Press, 2019.

Staudenmaier, John. *Technology's Storytellers: Reweaving the Human Fabric*. Cambridge, MA: MIT Press, 1989.

Steffen, Will, Wendy Broadgate, Lisa Deutsch, Owen Gaffney, and Cornelia Ludwig. "The Trajectory of the Anthropocene: The Great Acceleration." *Anthropocene Review* 2 (2015): 81–98.

Steinberg, Philip E. "Sovereignty, Territory, and the Mapping of Mobility: A View from the Outside." *Annals of the Association of American Geographers* 99, no. 3 (2009): 467–95.

Stokland, Håkon. "Field Studies in Absentia: Counting and Monitoring from a Distance as Technologies of Government in Norwegian Wolf Management (1960s–2010s)." *Journal of the History of Biology* 48, no. 1 (2015): 1–36.

Svärd, Per-Anders. *Problem Animals: A Critical Genealogy of Animal Cruelty and Animal Welfare in Swedish Politics 1844–1944*. PhD diss., Stockholm University, 2015.

Swan, James A. *In Defense of Hunting*. New York: Harper Collins, 1995.

Swanson, Heather Anne, Marianne Elisabeth Lien, and Gro B. Ween, eds. *Domestication Gone Wild: Politics and Practices of Multispecies Relations*. Durham, NC: Duke University Press, 2018.

Tally, Robert T. *Spatiality*. London: Routledge, 2013.

Taylor, Leighton. *Aquariums: Windows to Nature*. New York: Prentice Hall General Reference, 1993.

Taylor, Sunaura. *Beasts of Burden: Animal and Disability Liberation*. New York: New Press, 2017.

Thompson, Roger. *No Word for Wilderness: Italy's Grizzlies and the Race to Save the Rarest Bears on Earth*. Ashland, OR: Ashland Creek Press, 2018.

Thrift, Nigel. "Space: The Fundamental Stuff of Geography." In *Key Concepts in Geography*, edited by Sarah L. Holloway, Stephen P. Rice, and Gill Valentine, 95–107. London: Sage, 2003.

Tillhagen, Carl-Herman. *Allmogejakt i Sverige*. Stockholm: LTs Förlag, 1987.

Todd, Zoe. "Fish Pluralities: Human-Animal Relations and Sites of Engagement in Paulatuuq, Arctic Canada." *Études/ Inuit/ Studies* 38, no. 1–2 (2014): 217–38.

Tomes, Nancy. *The Gospel of Germs: Men, Women, and the Microbe in American Life*. Cambridge, MA: Harvard University Press, 1999.

Töpfer, Manuel. *Der Veterinärdienst des Deutschen Reiches in China von 1898 bis 1914*. Giessen: DVG, 2010.

Tortorici, Zeb. "Stereoscopic Animals: Spectatorship, Kodiak Bears, and the Keystone Animal Set." In *Zoo Studies: A New Humanities*, edited by Tracy McDonald and Daniel Vandersommers, 119–44. Montreal: McGill-Queens University Press, 2019.

Tsing, Anna Lowenhaupt. "Strathern beyond the Human: Testimony of a Spore." *Theory, Culture & Society* 31, no. 2–3 (2014): 221–41.

Tuan, Yi Fu. *Dominance and Affection: The Making of Pets*. New Haven, CT: Yale University Press, 1985.

Turkle, Sherry. *Reclaiming Conversation: The Power of Talk in a Digital Age*. New York: Penguin Books, 2015.

Urbanik, Julie. *Placing Animals: An Introduction to the Geography of Human-Animal Relations*. Lanham, MD: Rowman & Littlefield, 2012.

Urbanik, Julie, and Connie L. Johnston, eds. *Humans and Animals: A Geography of Coexistence*. Santa Barbara, CA: ABC Clio, 2017.

Valenze, Deborah. *Milk: A Local and Global History*. New Haven, CT: Yale University Press, 2012.

van Dooren, Thom, Eben Kirksey, and Ursula Münster. "Multispecies Studies: Cultivating Arts of Attentiveness." *Environmental Humanities* 8, no. 1 (2016): 1–22.

Veit, Helen Zoe. *Modern Food, Moral Food: Self-Control, Science, and the Rise of Modern American Eating in the Early Twentieth Century*. Chapel Hill: University of North Carolina Press, 2013.

Virilio, Paul. *The Vision Machine*. London: British Film Institute, 1994 [1988].

Vitebsky, Piers. "Centralized Decentralization: The Ethnography of Remote Reindeer Herders under Perestroika." *Cahiers du Monde Russe et Soviétique* 31, no. 2–3 (1990): 345–58.

von Essen, Erica. "The Impact of Modernization on Hunting Ethics: Emerging Taboos among Contemporary Swedish Hunters." *Human Dimensions of Wildlife* 23, no. 1 (2018): 21–38.

von Uexküll, Jakob. *Umwelt und Innenwelt der Tiere*. Berlin: Springer, 1909.

von Uexküll, Jakob, and Georg Kriszat. "A Stroll through the Worlds of Animals and Men: A Picture Book of Invisible Worlds." In *Instinctive Behavior: The Development of a Modern Concept*, edited by Claire H. Schiller, 5–80. New York: International University Press, 1957.

Warren, William. "Constructing an Econiche." In *Global Perspectives on the Ecology of Human-Machine Systems*, edited by John M. Flach, Peter A. Hancock, Jeff Caird, and Kim J. Vicente, 210–37. Hillsdale, NJ: Erlbaum, 1995.

Whitmarsh, Lorraine E. "The Benefits of Guide Dog Ownership." *Visual Impairment Research* 7, no. 1 (2005): 27–42.

Williams, Raymond. *Television: Technology and Cultural Form*. New York: Schocken Books, 1975.

Wischermann, Clemens, Aline Steinbrecher, and Philip Howell, eds. *Animal History in the Modern City: Exploring Liminality*. London: Bloomsbury Academic, 2019.

Wonders, Karen. "Habitat Dioramas and the Issue of Nativeness." *Landscape Research* 28, no. 1 (2003): 89–100.

Xu Zhanjiang 徐占江, Geriletu 格日乐图, Yang Ye 杨晔, Xu Xin 徐鑫, Zhao Yuxia 赵玉霞, and Tan Fujie 谭辅杰, eds. *Zhongguo Mo'ergele Ewenke ren* 中国莫尔格勒鄂温克人 [The Mergel Evenki of China]. Hulunbuir, China: Neimenggu wenhua chubanshe, 2013.

Xu Zhanjiang 徐占江, Zhao Yuxia 赵玉霞, Taolong 陶龙, Bi Jinjie 毕金杰, Wang Zhaoguo 王召国, Daxizhamusu 达喜扎木苏, Simujide 斯木吉德, and Sijidema 斯吉德玛, eds. *Zhongguo Buliyate Menggu ren* 中国布里亚特蒙古人 [The Buryat Mongols of China]. Hulunbuir, China: Neimenggu wenhua chubanshe, 2009.

Zaff, Brian. "Designing with Affordances in Mind." In *Global Perspectives on the Ecology of Human-Machine Systems*, edited by John Flach, Peter A. Hancock, Jeff Caird, and Kim Vicente, 121–56. Hillsdale, NJ: Erlbaum, 1995.

Zanolla, Serena. "Interactive and Multimodal Environments for Learning." PhD diss., University of Udine, 2013.

CONTRIBUTORS

ELLEN F. ARNOLD is a senior lecturer at The Ohio State University. A medievalist by training, in her work she focuses on the intersections of cultural, religious, and environmental history, and the history of water. She is the author of two monographs on medieval history, *Negotiating the Landscape: Environment and Monastic Identity in the Medieval Ardennes* (Penn, 2013) and *Medieval Riverscapes: Environment and Memory in Northwest* Europe (Cambridge, 2024). She has also been a longtime coeditor of the journal *Water History*, and is the author of the forthcoming *Water in World History* with Routledge.

CONCEPCIÓN CORTÉS ZULUETA is an independent scholar, with a PhD in art history. Between 2019 and 2024, she was a postdoctoral fellow at Universidad de Málaga, Spain. Her interdisciplinary research focuses on the presence and agency of nonhuman animals in contemporary art and audiovisual culture, addressing them as authors and creators. She leads the research project "Entomornithophilias (and Phobias): Impressions and Encounters of Birds and Insects," Plan Propio UMA. She has published several chapters and articles, coedited two books and, with Reyes Escalera, a special issue of *Boletín de Arte* on animals and art history. She has made research stays at La Sapienza, Università di Roma; the Universidade Nova de Lisboa; the National Art Library (Victoria & Albert Museum); Museo Nacional de Ciencias Naturales (Madrid); the New Zealand Centre for Human-Animal Studies (University of Canterbury); and the Massachusetts Institute of Technology, among others.

KARIN DIRKE is a professor in history of ideas at Stockholm University. Her work is situated within the environmental humanities and mainly concerns the complex dynamics of interspecies relationships in the modern period. She has published scholarly articles and book chapters exploring diverse topics such as human–large carnivore encounters, animal agency, farmed animals, and stable culture. Her particular fascination lies in elucidating and acknowledging animals' historical contributions, understanding them as integral agents in historical narratives.

AURORE DUMONT is an anthropologist. She is currently a fixed-term lecturer in Asian studies at École des hautes études en sciences sociales. Her research focuses on the pastoral and ritual practices among the Tungus and Mongol societies scattered in northeastern China. She seeks to understand how these ethnic minorities adapt their way of life to a fluctuating socioeconomic, religious, and political environment. Her recent publications include "Sacred Land Cairns and Shamanic Burial Sites on the Sino-Russo-Mongolian Borderlands," *Asian Ethnicity* 25 (2024); "Community, Faith and Politics: The Ovoo of the Shinehen Buryats throughout the 20th Century," *Études mongoles et sibériennes, centrasiatiques et tibétaines* 52 (2021); and "Turning Indigenous Sacred Sites into Intangible Heritage: Authority Figures and Ritual Appropriation in Inner Mongolia," *China Perspectives* 3 (2021).

CHARITY EDWARDS is a lecturer in architecture and urban planning and design at Monash University, where she directs the bachelor of architecture design studio program alongside teaching urban history and theory at undergraduate and master's level. Her research examines urban processes in remote and offworld environments that operate beyond the traditional container of the city, more recently via increasingly autonomous underwater technologies within the Southern Ocean. Her work has been published in *Dialogues in Human Geography*, *Society+Space*, and *Footprint: Delft Architecture Theory Journal*. Edwards is also a practicing architect with over twenty years' experience across Australia and Asia, and she continues to collaborate with artists, scientists, and communities to create buildings, objects, and landscapes.

AMELIA HINE is a critical resource geographer with the Marine Governance group at the Helmholtz Institute for Functional Marine Biodiversity, under the Alfred-Wegener-Institut and University of Oldenburg, Germany. Her research investigates the interplay of power and storytelling in resource and energy futures, with a focus on nonhuman and nonliving agencies and materialities, volumetric geographies, and particulate and chemical knowability.

DOLLY JØRGENSEN is a professor of history at University of Stavanger, Norway, specializing in histories of environment and technology. Her current research agenda focuses on cultural histories of animals, and her most recent monographs are *The Medieval Pig* (Boydell, 2024) and *Recovering Lost Species in the Modern Age: Histories of Longing and Belonging* (MIT Press, 2019). She is coeditor-in-chief of the journal *Environmental Humanities* and codirects the Greenhouse Center for Environmental Humanities at University of Stavanger.

FINN ARNE JØRGENSEN is a professor of environmental history at University of Stavanger, Norway, where he codirects the Greenhouse Center for Environmental Humanities. His research is placed at the intersection of technology and environment, drawing on history, digital humanities, media studies, and science and technology studies to explore technologically mediated relationships with nature. He is the author of *Making a Green Machine: The Infrastructure of Beverage Container Recycling* (Rutgers, 2011) and *Recycling* (MIT, 2019), and coeditor of several volumes, including *New Natures: Joining Environmental History with Science and Technology Studies* (Pittsburgh, 2013).

HETA LÄHDESMÄKI is a historian specializing in human-animal studies, human-wildlife conflicts, and conservation. She has studied, for instance, human-wolf relations in twentieth-century Finland and the entanglements between bird feeding and rat conflict in Helsinki. At the moment she works as a postdoctoral researcher at the University of Turku, Finland, in the Academy of Finland–funded HumBio-project, investigating the human relationship with the disappeared, endangered, introduced, and nonnative as well as invasive marine animals and plants in Finland.

RIIN MAGNUS is a research fellow in the Department of Semiotics, University of Tartu, Estonia, where she defended her dissertation, "The Semiotic Grounds of Animal Assistance: Sign Use of Guide Dogs and Their Visually Impaired Handlers." Her research focuses on ecosemiotics, urban environments, and relations between humans and nonhuman animals.

TATSUYA MITSUDA is an associate professor at Keio University, Yokohama, Japan. He was educated at Keio, Bonn, and Cambridge, where he completed a PhD in history. His research interests span the intertwined social and cultural histories of food, animals, and climate, with reference to the German and Japanese experience in the nineteenth and twentieth centuries. His most recent publications include "From Endurance to Escape: The Tokyo Summer as Lived Experience in the Twentieth Century," *Journal of Urban History* (forthcoming); "From Colonial Hoof to Metropolitan Table: The Imperial Biopolitics of Beef Provisioning in Colonial Korea," *Global Food History* 10 (2024); and "Consumed by Fear: Urban

Suspicions of Eating and Drinking in the Long Nineteenth Century," in *Handbook of Eating and Drinking: Interdisciplinary Perspectives* (Springer, 2024). He is currently finishing a book on the history of sweets and snacks in Japan and preparing a monograph on the history of animal health in nineteenth-century Germany.

TUOMAS RÄSÄNEN is professor of environmental history at University of Turku, Finland. He specializes in the environmental histories of wild animals, the history of environmentalism, and marine environmental history. Recently he has been studying people's changing perceptions of endangered animals, as well as their relationship with arthropods.

JENNY LEIGH SMITH is a historian and independent scholar based in western New York. She has published books and articles on the history of food security, famine, and the environmental impact of agricultural industrialization. She co-edited, with Tom Robertson, *Transplanting Modernity? The Environmental Legacy of International Development* (Pittsburgh, 2023).

NICOLE WELK-JOERGER is an interdisciplinary historian trained in art history, anthropology, and the history of science, technology, and medicine. She received her PhD in the history and sociology of science at the University of Pennsylvania and completed postdoctoral appointments at North Carolina State University and Princeton University.

INDEX